ANIMALS AS DOMESTICATES

THE ANIMAL TURN

ANIMALS AS DOMESTICATES

A World View through History

Juliet Clutton-Brock

Michigan State University Press

East Lansing

⊗ The paper used in this publication meets the minimum requirements of ANSI/NISO Z39.48-1992 (R 1997) (Permanence of Paper).

 Michigan State University Press
East Lansing, Michigan 48823-5245

Printed and bound in the United States of America.

18 17 16 15 14 13 12   1 2 3 4 5 6 7 8 9 10

LIBRARY OF CONGRESS CATALOGING-IN-PUBLICATION DATA

Clutton-Brock, Juliet.

Animals as domesticates : a world view through history / Juliet Clutton-Brock.
    p. cm. — (The animal turn)
    Includes bibliographical references and index.
    ISBN 978-1-61186-028-3 (case bound : alk. paper) ISBN 978-1-61186-064-1 (pbk : alk. paper) 1. Domestic animals—History. 2. Animals and civilization. 3. Human-animal relationships—History. I. Title.

    SF41.C577 2012
    636.08'2 –dc23   2011031274

Cover design by Erin Kirk New
Book design by Scribe Inc. (www.scribenet.com)

Cover art is of a caravan of Arctic reindeer herders on the move and is reproduced with permission from Øyvind Ravna's Photo Collection.

Back cover art is a seal from Mohenjo-Daro (2500 BCE) showing an Indian elephant and is used courtesy of Richard H. Meadow.

green press INITIATIVE Michigan State University Press is a member of the Green Press Initiative and is committed to developing and encouraging ecologically responsible publishing practices. For more information about the Green Press Initiative and the use of recycled paper in book publishing, please visit www.greenpressinitiative.org.

Visit Michigan State University Press at www.msupress.org

This book was written with the memory of all the animals who have been my companions over the past seventy years, beginning with my Rhodesian Ridgeback, Zimba, in 1941.

Contents

Foreword

JAMES A. SERPELL

DOMESTICATED ANIMALS ARE SUCH COMMONPLACE FEATURES OF THE MODERN WORLD that we tend to take them for granted. Yet evolutionarily speaking, this remarkably successful and ubiquitous category of living organisms is still a newcomer. A mere 20,000 years ago—a blink of the eye from the perspective of evolution—none of these animals existed, except perhaps for a few tamed wolves, soon to become dogs. At that time, the vast bulk of Earth's terrestrial vertebrate biomass consisted of wild animals, while the human population was a diminutive fraction of its current size, perhaps no more than a million persons altogether. Now, according to a recent report, domesticated livestock (including poultry) constitute a staggering 65 percent of all terrestrial vertebrates by weight.[1] Humans and their pets account for most of the rest, while "wildlife" represents a mere three percent of the total. In other words, within the space of a few thousand years, domesticated animals have virtually replaced their wild progenitors across most of the planet's surface.

In the process, domesticated animals have fundamentally altered the evolutionary trajectory of our own species. In terms of its influence on human culture and ecology, animal domestication easily rivals the invention of stone tools, religion, written language, mathematics, and the more recent industrial and information technology revolutions. But it receives far less attention and credit from scholars than might be expected given its extraordinary impact. After roughly a million years of living successfully as hunters and gatherers, humans embarked on a radical, unparalleled, and unprecedented shift in lifestyles, with the enfolding of other species of living animals within their societies. The domestication of animals developed, the author believes, from the unique nurturing behavior of the human species, and once established it soon spread across the globe, leaving few, if any, ecosystems and cultures unaffected. So, what caused this departure from nomadic hunting and gathering? Where and when did it happen, and why? And why were only some species domesticated and not others? These and other fundamental questions form the framework for this wonderfully illustrated and informative book.

Juliet Clutton-Brock is a world-renowned zooarchaeologist and probably the greatest living expert on the history and prehistory of domesticated animals. She was already an established authority in the field when I was taking undergraduate courses in archaeological science back in the early 1970s, and certainly nobody is better qualified to write a book on this complex and fascinating topic. At the same time, Clutton-Brock knows how to tell a story. She writes in a clear, accessible, and down-to-earth style and possesses a remarkable ability to conjure up evocative images of the past from the scattered traces left by ancient remains and artifacts. She also reveals a true affection for her subject matter. *Animals as Domesticates* is not only an outstanding synthesis of archaeological, scientific, literary, and pictorial evidence, but it is also an engaging celebration of the extraordinary diversity, adaptability, and resilience of all those dogs, cats, cows, sheep, goats, horses, donkeys, camels, chickens, turkeys, rabbits, and so forth, to whom the author has dedicated her life's work.

Acknowledgments

My first and greatest thanks go to Linda Kalof (Michigan State University), who invited me to write this book and who has overseen its production and dealt with countless e-mails. I am indebted to Seven Bryant (Michigan State University) for obtaining all the copyright permissions for the illustrations and for sorting out the ensuing complications. My thanks are due to Richard Meadow (director of the Zooarchaeology Laboratory, Peabody Museum, Harvard University) for illustrations and for his comments on the draft of chapter 7. I would like to thank James Serpell (University of Pennsylvania) for writing his very supportive foreword. Finally, I thank those authors and institutions that have allowed their illustrations to be reproduced without charge: Agnes Arnold-Forster, Joel Berger, Louis Chaix, Joost Crouwel, Galen Frysinger, Colin Groves, Rebecca Jewell, Richard Meadow, Steven Mithen, Joan Oates, Bob Partridge, Julian Thomas, Laverne Waddington, Ya-Ping Zhang, Amerind Foundation, Free World Academy, Hungarian National Museum, Kenyon Institute, and the University of KwaZulu-Natal Press.

Introduction

Aristotle, who was born in 384 bce, believed that everything in nature had a purpose, and this purpose was for the benefit of mankind. He wrote, "plants are evidently for the sake of animals, and animals for the sake of Man; thus Nature, which does nothing in vain, has made all things for the sake of Man."[1] For more than 2,000 years, from writings that are even earlier than those of Aristotle until the discovery of the laws of evolution, it was usual throughout the Western world to believe that the universe had been created in a "Scale of Nature" or a "Great Chain of Being" with "man" at its pinnacle.[2] The widespread view that the world is divided into the "human," which is inherently unnatural, and the "animal," which is natural, is the cultural inheritance from this historical premise and from the belief that humans are created in the image of God and have an everlasting soul, while animals are without souls.

Today, the great majority of people in the Western world still believe that animals and plants exist for their benefit, and this attitude is unlikely to change, although the belief that humans are the only beings to have a soul and therefore that they are above the rest of creation is losing credence. It is becoming slowly more commonplace for Homo sapiens to be included as a part of the animal world and even to be described and classified as a close cousin of the chimpanzee.[3] The beginning of this fundamental change, which has affected so many social attitudes to religion, science, politics, and philosophy, may be attributed to Thomas Henry Huxley in his essays of 1863 in *Man's Place in Nature*.[4] The essays were written at a time when this was a topic of great concern and debate among every level of society, following the publication of Charles Darwin's *On the Origin of Species* in 1859 but before his *The Variation of Animals and Plants under Domestication* in 1869.[5]

Homo sapiens evolved as "man the hunter,"[6] and for by far the greater part of their early history the relationship of humans with other animals would have consisted of straightforward predation. In early hominins this predation was probably comparable with the hunting in the wild today by clans of living chimpanzees on other species of animals, which they kill for their meat. In the more recent Neanderthals and anatomically modern humans during the last Ice Age, the social hunting of large mammals such as mammoths or bison would have more closely paralleled that of wolf packs, and indeed the first partnership between wolf and humans may have begun from their joint hunting skills at least 30,000 years ago. A notable difference, however, between wolf and human hunters was, and is, that in wolves, an alpha female is usually the pack leader, while human hunters are usually led by a man.

The human species has evolved what appears to be a unique capacity for wanting to tame, nurture, and live with almost all species of vertebrate animals. Often this begins with the "cute response" that is elicited by the baby face of young animals, and a rational explanation for why humans are inherently attracted not only to human babies in general but also to baby animals may be looked for from the work of Sarah Blaffer Hrdy in her recently published *Mothers and*

Others. Hrdy describes how humans, alone among the great apes, are communal breeders.[7] They readily nurture each other's children, and indeed without the active help of other members in hunter-gatherer societies, she claims that few children would survive to adulthood. Among all species of primates, human babies are the largest, latest maturing, nutritionally costliest offspring, and yet humans breed the fastest, with an average birth interval of two to three years, while other great apes average a birth interval of six years. Hrdy accounts for this with the human capacity for shared care. Furthermore, she believes that the "seesaw" of enlisting care and giving care created the impetus for language and therefore the evolution of bigger brains and the ability for empathy with the emotions and desires of others. Following Hrdy's thesis on the evolution of Homo sapiens as communal breeders, it is easy to see how hunter-gatherers could expand the shared care of their infants into the enfolding of young animals into their social groups. If communal breeding and social hunting are the prime behavioral patterns of Homo sapiens, then the desire to enfold other species of animals within human societies may be explained as having evolved from the combined instincts for nurture and domination.

The instinct for humans to nurture was well understood in the nineteenth century by Darwin's cousin Francis Galton, who published a far-seeing essay titled the "Domestication of Animals" in 1883.[8] Galton has become ill-famed for his theory of eugenics—that is, the improvement of a human population by selective reproduction, which he developed from Darwin's theory of the origin of species by natural selection. Despite his unpopular theories, like Darwin and many scientists of his day, Galton was a polymath whose inquiries ranged over a huge number of interests, and his essay leaves little more to be said on the "why and wherefore" of domestication.

Galton could be described as an anthropologist and psychologist who, for his time, had a rare understanding of animal behavior and the human mind. In quoting what he claims is the earliest written description of a pet, he gives a perfect example of the human instinct for nurture.[9] It is in the parable in the Old Testament spoken by the prophet Samuel to King David: "The poor man had nothing save one little ewe lamb, which he had bought and nourished up: and it grew up together with him and with his children; it did eat of his own meat, and drank of his own cup, and lay in his bosom, and was to him as a daughter."[10] In describing the dog "as the animal above all others the companion of man," Galton wrote: "As the man understands the thoughts of the dog, so the dog understands the thoughts of the man, by attending to his natural voice, his countenance, and his actions. A man irritates a dog by an ordinary laugh, he frightens him by an angry look, or he calms him by a kindly bearing. . . . Who for instance, ever succeeded in frowning away a mosquito, or in pacifying an angry wasp by a smile?"[11] Galton may not have considered that insects would be likely pets, but singing crickets are traditionally kept in little cane cages by the Chinese,[12] and the daughter of the Victorian writer Edmund Gosse, a near contemporary of Galton, is recorded as having kept a blowfly as a pet in a cage until it sadly died of "a surfeit of sugar."[13]

Galton ends his essay on the domestication of animals with this proverbial statement: "It would appear that every wild animal has had its chance of being domesticated, that those few which fulfilled the above conditions were domesticated long ago, but that the large remainder, who fail sometimes in only one small particular, are destined to perpetual wildness so long as their race continues. As civilization extends they are doomed to be gradually destroyed off the face of the earth as useless consumers of cultivated produce."[14]

THE BIOLOGICAL PROCESS OF DOMESTICATION

Where Galton's essay was essentially on the power of nature over nurture, and he correctly saw this as the reason why so few species of animals have succumbed to full domestication, the fundamental process by which wild species are changed into domestic forms had already been published by Charles Darwin in 1868. Darwin realized that it is the principle of both natural and artificial selection that underlies the evolution of domestic animals and plants, and there is little new that can be added to his description of the three kinds of selection:

> The principle of selection may be conveniently divided into three kinds. *Methodical selection* is that which guides a man who systematically endeavours to modify a breed according to some prede-termined standard. *Unconscious selection* is that which follows from men naturally preserving the most valued and destroying the less valued individuals, without any thought of altering the breed; and undoubtedly this process slowly works great changes. Unconscious selection graduates into methodical, and only extreme cases can be distinctly separated; for he who preserves a useful or per-fect animal will generally breed from it with the hope of getting offspring of the same character; but as long as he has not a predetermined purpose to improve the breed, he may be said to be selecting unconsciously. Lastly, we have *Natural selection* which implies that the individuals which are best fitted for the complex, and in the course of ages changing conditions to which they are exposed, generally survive and procreate their kind. With domestic productions, natural selection comes to a certain extent into action, independently of, and even in opposition to, the will of man.[15]

Darwin's term "methodical selection" has since become known as "artificial selection," and it is the means by which the thousands of breeds of domestic animals and plants all over the world have been produced and continue to be developed. To understand how artificial selec-tion works, it is necessary to define domestication and describe the biological process that has enabled, for example, the wolf to become our most highly valued animal companion, the dog.

An animal may be said to be domestic when it has lost its fear of humans and will breed in captivity, but true domestication involves much more than this and can be defined as the keeping of animals in captivity by a human community that maintains total control over their breeding, organization of territory, and food supply.[16]

True domestication takes place from the combination of a biological process and a cultural process. The biological process begins when a few animals are separated from the wild species and are tamed—that is, they become habituated to humans. A few of these animals may breed within the human community, and if their offspring survive, they will form a small isolated or founder group. If these animals then interbreed and increase in number over many successive generations, they will respond by means of small genetic changes to natural selection under the new regime of the human community and its environment, and later they will respond for economic, cultural, or aesthetic reasons first to Darwin's unconscious selection and then to artificial selection until the domestic breed is created. The breed may be defined as a group of animals that has been bred by humans to possess uniform characters that are passed down through the generations and distinguish the group from other animals within the same species.

Once domestication has occurred and a new species is established, say the domestic dog, new breeds are produced from the domestic form by further reproductive isolation leading to genetic drift, as in the founder populations of new subspecies in the wild. The founders of the new breed contain only a small fraction of the total variation of the parent species, and it becomes a genetically unique population, which continues to evolve under natural and artificial selection. At any point the process can begin again, and further new breeds can be developed by crossbreeding.

Over the course of time, natural selection plays a considerable role in the production of breeds, and climate is a determining factor in their evolution. The effects of climatic selection on domestic animals appears to be identical to the well-known correlations in size and body shape that can be seen in subspecies of wild animals across a geographical cline. This can be seen, for example, in the domestic cat, which in the cold climate of northern Europe has a heavy body and head, short ears, and relatively short legs, while in the hotter and drier Mediterranean region and farther south, domestic cats are lanky, with big ears and long legs like their presumed ancestor, the African wild cat, *Felis silvestris lybica*.

Exactly the same process can be seen in the development of breeds of domestic horses. In the cold North, the horses are traditionally small, stocky, and heavy-headed, while the Arabian horse, with its fine head and long, slender legs, evolved in hot, arid regions.

Domestication can be seen as an evolutionary process that mimics the sequence of events when a small group of wild animals becomes isolated from the main group, say on an island. The subsequent evolution of this tiny founder group will depend on its chance genetic composition. The principle of how this works is known as the founder effect, and it will cause the original and the new, isolated populations to evolve along different paths. Initially the founder effect will lead to a population bottleneck caused by fluctuations in the gene frequencies, which is known as genetic drift. The population contracts and then expands again by breeding, but with an altered genetic composition. Genetic drift and natural selection act together in the new population of animals, and after thousands or even millions of years a new species will evolve, as did the well-known example of Darwin's finches on the Galapagos Islands.[17]

Before true speciation occurs in the isolated group of wild animals there may be a halfway stage, which is termed the "subspecies." A subspecies is a geographically and reproductively isolated group of animals or plants that differs in hereditary characteristics from the main species but is not distinctive enough to be treated as a separate species. Isolation is the important factor because it prevents interbreeding and hence a flow of genes between the populations. The subspecies is often a stepping stone to true speciation, and in the same way, the remains of animals that appear to be halfway between the wild and the domestic are found on early prehistoric sites.

If the isolated group consists of tamed animals that are under human control, then unconscious and artificial selection will also be a powerful force in the progeny, and the process of change will be speeded up. The question then has to be asked whether the new, isolated group of humanly controlled animals can be considered to be a new species or subspecies. What constitutes a species of animals? A common definition of a species is: if a population of animals breeds freely and produces fertile offspring, then it is a species; if two populations do not interbreed or if the hybrid offspring are infertile, then they are separate species—for example, the horse and the donkey. However, many animals that are typically considered to be good species will interbreed with fertile offspring, although, because of a behavioral or geographical barrier, they may not usually do so in the wild—for example, the wolf (*Canis lupus*) and the coyote (*Canis latrans*).

Another definition of species that avoids whether or not the animals will be fertile when interbred has therefore become necessary, and it is known as the biological species concept. Proposed by the great biologist Ernst Mayr as long ago as 1940, the concept is that a species is a group of actually or potentially interbreeding natural populations that is reproductively isolated from other such groups.[18] This leads to the question of whether domesticated animals should be considered to be separate species from their wild progenitors, which has been debated since the time of Darwin. All domestic animals will interbreed with their wild progenitors, so in that sense they are not separate species, but, on the other hand, they are reproductively isolated from them. A biologically more important factor is that molecular evidence is now proving that a number of domestic forms—for example, domestic sheep—are descended from several wild lineages, so they should not be aligned with one particular ancestor and should be treated as separate species in their own right.

For the past fifty years or more there has been discussion among archaeozoologists about whether the names of domestic animals should be included in the Latin nomenclature that is used for all wild species. It is usual for scientists to consider domestic animals as artifacts of human design and therefore as unnatural products of human manipulation. This has led to a movement to include the names of domestic forms as part of the name for the wild species—for example, *Canis lupus f*(orma) *familiaris*.[19] However, if looked at logically, it can be seen that classifying the dog as a wolf, and not accepting domestication as an evolutionary process that has led to the production of new species, is for cultural and historical reasons rather than for biological ones, for who could deny that a breed of dog like the spaniel, which has been isolated from its progenitor, the wolf, for thousands of years, fits the biological species concept? The dog should therefore be given a different Latin name from the wolf, and the most parsimonious arrangement is to continue to call the dog and all other fully domestic animals by the Latin names they were first given by Linnaeus in the eighteenth century, which for the dog was *Canis familiaris*. These Latin names are used throughout this book and are listed in the Appendix together with the names that are now accepted by the International Commission on Zoological Nomenclature (ICZN) for the wild ancestral species.[20] There are, in addition, some species that have the same Latin name in the domestic as in the wild species—for example, the reindeer, *Rangifer tarandus*. The Latin names are important because they give an individual identity to the domestic species, and, at the same time, they indicate the degree of relationship to other species. For example, sheep with the name *Ovis aries* and goats with the name *Capra hircus* often look very similar and live in the same environments, but their Latin names show their taxonomic separation and the fact that they will not interbreed.

Domestic animals that return to living in the wild are described as feral, and can be defined as populations that live in a self-sustained population after a history of domestication. Examples are the "wild" horses of the Great Basin.[21] Feral populations usually retain the Latin name of the domestic species, although there are exceptions, the most notable of which is the dingo, which is assumed from biogeographical and genetic evidence to have been taken to Australia by humans. However, its history is so ancient and dingoes have lived as wild populations for so long that, by common usage, they are treated as a separate species from the domestic dog and given the name *Canis dingo* rather than *Canis familiaris*.

THE CULTURAL PROCESS OF DOMESTICATION

The term "culture" has had different meanings in different periods of history and depending on the part of human society to which it is being applied. In the eighteenth century the term was applied to the "improvement" by selective breeding of plants or animals. Later, a "cultured person" was one who was highly educated and "refined." In the twentieth century "culture" came into use by anthropologists for the particular way of life of human societies in different parts of the world, and this is the sense in which it is now used for the learned behavioral patterns of animal societies, for just as humans can no longer be separated from animals on the basis of "man the toolmaker" as was common in the 1960s,[22] so it is argued that it is no longer justified to say that only humans have culture.[23] Many groups, or societies, of wild animals have been proved to use simple tools, most notably chimpanzees but also birds, particularly crows and finches, and within the past decade it is becoming accepted that animal societies have traditions of learned behavior that can be described as "culture." Frans de Waal gives this definition of culture:

> Culture is a way of life shared by the members of one group but not necessarily with the members of other groups of the same species. It covers knowledge, habits, and skills, including underlying tendencies and preferences, derived from exposure to and learning from others. Whenever systematic variation in knowledge, habits, and skills between groups cannot be attributed to genetic or ecological factors, it is probably cultural.[24]

Tim Ingold has argued that there is a relationship of mutual trust between hunter-gatherers and their wild prey in which the environment and its resources are shared, but with domestication the relationship changes to total human control and domination.[25] In order for animals that have been prey to be domesticated, they have to be incorporated into the social structure of the human community and become objects of ownership, inheritance, purchase, and exchange. The animal is removed from the wild, where it learns from birth either to hunt or to flee on sight from any potential predator; the tamed animal is brought into a protected place, where it has to learn a whole new set of social relationships as well as new feeding and reproductive strategies, and under domestication, this "culture" is passed down from generation to generation.

In the anthropological sense, as used by Frans de Waal and then applied to the domestic animal, culture may be defined as a way of life imposed over successive generations on a society of humans or animals by its elders. Where the society includes both humans and animals then the humans act as the elders.

A domestic animal is a cultural artifact of human society, but it also has its own culture, which can develop, say in a cow, either as part of the society of nomadic pastoralists or as a unit in a factory farm. Domestic animals live in many of the same diverse cultures as humans, and their learned behavior has to be responsive to a great range of different ways of life. In fact, so closely do many domestic animals fit with human cultures that they seem to have lost all links with their wild progenitors. The more social or gregarious in their natural behavioral patterns are these progenitors, the more versatile will be the domesticates, with the dog being the earliest animal to be domesticated more than 15,000 years ago. The dog is an extreme example of an animal whose culture has become humanized even to the extent that credence has been given by experiment to the well-known maxim that dogs and their owners often look alike.[26]

THE IMPETUS FOR DOMESTICATION

It is not fully understood why the widespread domestication of livestock animals, these being sheep, goats, cattle, pigs, and equids in the Old World and camelids in South America, occurred progressively from 8,000 years ago, but this was the basis of the so-called Neolithic revolution when the fundamental change in human societies occurred and groups of nomadic hunter-gatherers began to live in settlements and became farmers and stockbreeders. Archaeologists in the past have hypothesized that there was a natural progression first from generalized or broad-spectrum hunting in the Palaeolithic, at the end of the last Ice Age, to specialized hunting and herd following, say of reindeer or llama. It was then believed that this stage was followed by control and management of the herds, then by controlled breeding, and finally by artificial selection for favored characteristics. However, the sequence would very rarely have been so smooth, for the social implications of ownership by a social group of hunter-gatherers are a bigger hurdle to domestication than they may seem. Many hunter-gatherer societies that could have domesticated animals never did so, and this was probably for cultural as much as for many other complicated reasons. Why, for example, were caribou (*Rangifer tarandus*) never domesticated by the native peoples of Alaska and Canada, while reindeer, which belong to the same species, have been domesticated in Arctic Europe for thousands of years?[27]

In 1992 Stephen Budiansky published a fresh look at the discussions on how domestication came about. He intended his thesis to be contentious and to be a wake-up call for all the archaeologists and biologists who for generations had claimed that domestication of the dog and livestock animals came about through taming of captured animals. He argued that if this is how it happened, then why had not all the other animals that had been tamed by the ancient Egyptians and others also not been domesticated? His answer was that domesticated species were those that had themselves contributed to the process by a combination of being opportunistic individuals that had learned to glean their food from around human settlements and then in future generations had become submissive through the retention of their juvenile development in the absence of predators.[28]

Francis Galton had written what may be considered a more scientific explanation of why so few species of animals have been domesticated in his essay of nearly 100 years earlier, and he was followed by Jared Diamond in his popular account of domestication.[29] Galton realized that the natural behavior patterns of the wild species were the controlling factor for the ease with which species of animals could be enfolded into human societies, although as everyone knows who lives close to animals, each individual animal has its own personality, and one will be submissive while another will be aggressive. The popularity of Budiansky's theories has, however, led to a useful, widening of discussion about the origins of domestication and the role played by the animals themselves in the different stages from their first association with humans to their final exploitation in the factory farm.

An animal can be tamed only after there has been a breakdown of the barriers of fear, for lack of fear is the essence of tameness, but the extent of this fearlessness depends on the complex characteristics of the species as well as the individual temperament of the animal. An animal will be more or less fearful, and therefore more or less easy to tame, depending on whether it belongs to a social or solitary species, whether it is normally a predator or prey, and, if it is a prey, whether it escapes by fleeing or hiding. A burrowing species like a rabbit can be easily tamed, while a superficially similar species like a hare that depends on speed for escape cannot be tamed. Similarly, a wild goat, which is by nature a mountain animal that depends on agility among its rocky environment for escape, has less instinctive fear than a gazelle, which is a plains animal dependant on speed for escape from predators. This means that the natural flight distance of a goat is much shorter than that of a gazelle and partly explains why goats could be easily tamed and then domesticated by prehistoric people, while gazelle remained wild.[30]

The reason why the dog was the first animal to have been enfolded into human societies and why dogs are the most companionable of all animals is because the natural behavioral patterns of their progenitor, the wolf, are so like those of humans. First and foremost, wolves live in a hierarchical society with a pack leader, and they have an element of active submission in their genetic constitution, so it is natural behavior for a dog to join its owner's family or group and to accept the dominance of one member.

The physical and emotional changes that are brought about by the process of domestication need only be summarized here as they have been well documented elsewhere, at least in mammals.[31] An imbalance and disruption in the rate of growth of different parts of the body lead particularly to reduction in overall size. In cattle, sheep, and goats, this leads to reduction in horn size and shape,[32] and in carnivores to reduction in the length of the jaws, causing compaction of the cheek teeth. These changes enable the archaeozoologist to identify the skeletal remains of the earliest domestic animals when they occur on archaeological sites. Many of the characters originate from the retention of juvenile development into adult life, and they have much the same form in species as different as the dog and pig—for example, lop ears, change of coat color, and a curled tail. Internal changes that are widespread in mammals are in the increase in the layers of fat and a reduction in the size of the brain, which can be measured in the cranial capacity of the skull.

All of the above physical characters were produced in the well-known experiments by Belyaev and Trut, who over forty years selected a population of 45,000 silver foxes for only one behavioral character: tameness.[33] Selection for this single behavioral trait also altered the physical constitution and appearance of the animals, thereby showing that the changed characters were almost certainly not individually selected for in the earliest stages of domestication. They are an interconnected suite of mainly juvenile characters controlled by endocrinology, physiology, morphology, and behavior.[34] With the recent great increase in the science of genetics, biologists are studying the molecular basis for domestication with searches for genes specifically responsible for each of the common traits, such as coat color, lop ears, and so forth.

An aid to the domestication process that does not require selection is castration, which leads to great changes in male animals. It causes the bones to continue to grow in length but not in girth, and it increases the deposits of body fat. An ox will have a heavier body but longer, more slender limbs and horns than a bull fed on the same diet. But most important for the human owner, castration turns an aggressive bull into a placid and submissive plow ox. It is probable that castration has been carried out on male animals since the beginning of livestock husbandry.

Throughout the twentieth century, archaeozoologists studied the history and relationships of wild and domestic animals by direct investigation of skeletal remains from archaeological sites of all periods and from all regions, and much could be learned from the morphology and measurement of bones. During the present century, however, molecular biology, isotope studies, and radiocarbon dating now commonly supplement osteology in providing extraordinary details of the near absolute date of death, the genetic history and taxonomic relationships, and even the diet of individual animals (or humans) from fragments of bone.

The following chapters are arranged by region and by broad period, and descriptions are given of the early history of the indigenous and introduced domestic species from the time of their first enfoldment into human societies. Over the millennia, humans have lived in a symbiotic relationship with a wide diversity of animals, and their partnerships have been centered on much more than physical resources. The final chapter points to the changes that have occurred in these partnerships in modern times. The huge increase in the human population of the world is cited with the effects that the correspondingly huge increase in food, leather, and all the necessities for modern life have had on farming. The personality of each individual livestock animal is lost sight of in factory farms, where rows of caged animals are treated like animate vegetables. Meanwhile, ethologists and geneticists are studying the minds of animals and finding that in consciousness, memory, and morality, their closeness to the human mind would surprise many of these farmers.

Eurasia after the Ice

Around 20,000 years ago the Northern Hemisphere was in the grip of the coldest phase of the last Ice Age, the Neanderthal race of humans (*Homo neanderthalensis*) was almost extinct, and anatomically modern humans (*Homo sapiens*) were living as hunter-gatherers in small groups wherever the icy cold allowed them to find food and shelter. But the world was warming up, and in "fits and starts" over the next 5,000 years the huge ice sheets that covered the land and sea were beginning to melt. Between 17,000 and 13,000 years ago a few families in western Russia and central Europe must have been keeping tame wolf pups because the remains of canids that are morphologically different from the bones and teeth of wolves begin to appear on archaeological sites, along with the skeletal remains of wild horses and other animals that had been hunted for food.[1]

The archaeozoological evidence indicates that wolves were the first species to be domesticated, and it is not difficult to see why, for the gray wolf (*Canis lupus*), progenitor of all dogs, is a ubiquitous species that was originally distributed over the entire Northern Hemisphere. Wolves can flourish in lands that vary from the deserts of Arabia to the Arctic tundra. While Eurasia was still covered in ice sheets, wolves that became dogs, being carnivores and scavengers, could live off the detritus around the temporary camps of nomadic human hunters, even where the winter temperature reached minus 30 degrees Celsius or even lower, and they could travel over any distances in a symbiotic relationship with their human companions.

With the advance of molecular science there have been several major studies of the genetic history of dogs and the relationships of breeds. One of the earliest of these, by Carles Vilà and colleagues in 1997, caused much controversy with its premise that the genetic separation between wolf and dog occurred around 135,000 years ago, which is before the emergence of anatomically modern humans.[2] Until recently there has been no morphological evidence for the presence of a separate canid that could be described as "dog" before around 17,000 years ago. However, a multidisciplinary study of canid remains from Palaeolithic sites in Belgium, Ukraine, and Russia appears to show that there were "dogs" that could be distinguished from wolves living on some sites 30,000 years ago.[3] The most spectacular canid find from these sites is, however, later in date. It is from the Epigravettian Eliseevich I site on the Russian plain, and has an age of around 13,900 years. There, a dog skull, one of two, was found in a hearth deposit near a concentration of mammoth skulls. Its braincase had been perforated on the left and right sides, and cut marks are present on the zygomatic and frontal bones. With the exception of the canines and some premolars, all its teeth are missing. In addition, the left and right carnassial teeth were apparently removed by cutting into the alveoli.[4] It should be noted, however, that even if it can be proved with some certainty from morphological and genetic analyses that there were "dogs" in the Palaeolithic 30,000 years ago, this is still 100,000 years after the claim of Vilà and colleagues for a genetic divergence between wolf and dog at the very early date of 135,000 years ago.[5]

The problem with believing in the establishment of a race of canid evolved from but genetically separate and morphologically distinct from the wolf is that the "dogs" would have to be reproductively isolated over many generations from their progenitor, the wolf. It is difficult to see how this could be brought about with the small number of "dogs" that would be living as commensals with nomadic hunters in the Ice Age tundra. When a bitch came into oestrus, she would be mated by the local wolves, and her puppies would not retain the changed behavioral patterns of tameness and lack of aggression that enabled their mother to live in close association with a human group.

The coldest phase of the last Ice Age was between 20,000 and 12,000 years ago when the tundra of northern Europe and Asia was dominated by the "mammoth fauna." This is the name given to the assemblage of large mammals whose remains are commonly identified from geological and archaeological deposits of this period. The main species besides mammoths were bison, rhinoceros, horse, and reindeer, and it was these mammals that were painted with such wonderful artistry on the rock walls of caves in what is now France and Spain. Besides this megafauna there were many smaller mammals such as wolf, boar, wolverine, hyena, and pika, but so far no paintings of "dogs" have been identified.[6]

At the Last Glacial Maximum, around 20,000 years ago, small groups of human hunters and their families were eking out a living in the freezing cold of the tundra. The remains of their habitation sites have been excavated in the Ukraine, from where the remains of their shelters have been found, built out of mammoth bones, tusks, and hides.[7] In more southerly parts of Europe the hunters were living a nomadic existence following herds of wild horses and reindeer. The site of the Roche de Solutré in the southern part of Burgundy in France is famous for its skeletal remains of vast numbers of wild horses, but also of bison and reindeer that were killed by Palaeolithic hunters between around 55,000 and 12,000 years ago. It was originally believed that herds of horses must have been driven to their deaths over the rocky cliff, but recent interpreters suggest it is more likely that small bands of horses were either

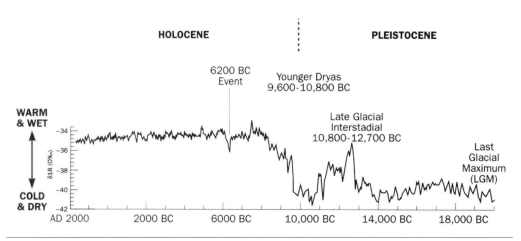

Figure 1. Fluctuations in the global temperature from around 20,000 years ago until the present day. (Source: Steven Mithen, *After the Ice: A Global Human History, 20,000–5000 BC* [London: Weidenfeld and Nicolsen, 2003], 12. Copyright 2003 Steven Mithen; reproduced with permission.)

driven up to the rock and slaughtered there or were driven into a corrallike enclosure.[8] In the later period it is quite probable that the hunters were helped in the drives by dogs. No skeletal remains of dogs have been excavated from the site, but then it is unlikely that dogs would have died at the rock, which was a specialized place of slaughter for horses.

MEDITERRANEAN ISLAND FAUNA IN THE UPPER PLEISTOCENE

During the Upper Pleistocene, southern Europe and the Mediterranean islands lay outside the zone of extreme cold. These islands had also been isolated in the deep Mediterranean Sea for millions of years, and many had evolved endemic species of animals and plants that were markedly different from their ancient ancestors on the mainland of Africa, Asia, and Europe.[9] The best known of the animals were the dwarf hippos and dwarf elephants, whose skeletal remains have been commonly found on Sicily, Sardinia, Crete, Cyprus, Malta, and Rhodes. The dwarfism of the several species of elephants and hippos, identified from their fossil remains, is assumed to have evolved from natural selection due to the restricted diet and lack of predators.

On Sardinia, there was an endemic jackallike canid, and great numbers of an endemic and unique goatlike herbivore, named *Myotragus balearicus*, were living on the Balearic Islands. These may have survived until the first people reached Mallorca and Minorca in the Neolithic. William Waldren, excavator of the cave sites where great numbers of bones and horn cores of this strange "goat" were found, believed that the Neolithic immigrants had made some attempt to domesticate the animals.[10] However, if this endemic "goat" did survive until around 2000 BCE (suggested by radiocarbon dating),[11] this late date was unlike that for all the other endemic species of mammals on the Mediterranean islands in the Upper Pleistocene. These all appear to have become extinct soon after the great warm-up that began 12,000 years ago.[12]

The great question is: did the extinctions occur as a result of the dramatic climate change, or were they caused by the first human hunters to arrive on the islands? Neither radiocarbon dating nor archaeozoology has so far succeeded in answering this question, nor have they fully done so for the extinction of the mammoth fauna of the whole Northern Hemisphere. There is, however, one cultural legacy to survive from the fossil skulls of the dwarf elephants that have been frequently found on the islands since the beginnings of Greek writing. The skull of an elephant has a single large round cavity in the frontal bone, the nasal orifice, to which the muscles of the trunk are attached, and it is now generally believed that this cavity gave rise to the legend of the land of the Cyclops, the one-eyed giants in Homer's *Odyssey*.[13] In about 800 BCE, when writing in ancient Greek first appears, and when it is assumed that Homer's *Odyssey* and *Iliad* were first written down, the skulls of dwarf elephants would often have been found in the island caves, and the legend would easily have been spread throughout ancient Greece that they were the skulls of one-eyed giants.[14]

There is no evidence of human beings in Britain at the Glacial Maximum or for the following 5,000 years while northern Europe began to warm up. The first hunters arrived at about 12700 BCE when Britain and Ireland were still very much a part of northern Europe, with an encircling coastline that was joined to Scandinavia. This meant that there was free movement of animals and people for hundreds of miles from south to north once the ice had gone.[15] However, the rapid growth and spread of deciduous forests soon created a new barrier. After the final cold snap of the Ice Age, in the period known as the Younger Dryas, the temperature rose by 7 degrees Celsius, and by 10,000 years ago, dated by pollen analysis and radiocarbon, Europe entered the Preboreal phase for 1,000 years. This was followed by the Boreal phase of the current epoch, the Holocene, from 9,000 to 8,000 years ago. By the end of the Boreal the melting ice had caused dramatic rises in sea level, and Ireland had been cut off from Britain and Europe for some hundreds of years. The English Channel broke through at the end of the Boreal, and the North Sea cut off the east coast from northern Europe, which meant that the people, animals, and plants that had moved into Britain became isolated.

Over the whole of Europe by this time, the Ice Age megafauna were extinct or driven to the far north of Europe and Asia. A small population of mammoths had clung on in a dwarf form on Wrangel Island, north of Siberia (extinct by 4,000 years ago), while reindeer had retreated to where they could feed on the mosses and lichens that are their specialized diet. Herds of wild horses still roamed over open land, but probably not in the huge numbers that grazed on the Ice Age tundra. However, the lands of Europe and Asia were not empty of large mammals; there was a replacement fauna of wild cattle, aurochs, several species of deer, wild boar, bears, and wolves, as well as the wild horses and a new species of forest bison.[16] There were no species of sheep or goats in Europe, but several in Asia.[17] All these species of mammals were soon to have their populations drastically altered in numbers and distribution by human impact, either by burning of the landscape, hunting, or domestication.

The dramatic change in the large mammal fauna at the end of the Ice Age was followed by replacement of the Palaeolithic nomadic hunters by Mesolithic hunter-gatherers who began to live a settled existence and began to cultivate plants and enfold animals into their societies all over the Northern Hemisphere. The Palaeolithic hunters' primary method of killing large prey had been by short-range stabbing with large stone axes. This also changed with the Mesolithic hunters, who used long-range missiles in the form of bows and arrows that were tipped with small, very sharp flint flakes (microliths). It very probably became more and more common for the hunters to use dogs to track the wounded prey.[18]

During the early Mesolithic in Europe, around 9,000 years ago, dogs were the only species of animal whose remains show osteological evidence of human association from their small size and compacted teeth in the jawbones. They must have been a rare accompaniment to the human settlements, which were still only temporary camps. One of the earliest Mesolithic dogs has been identified from a complete skull excavated from the Early Mesolithic site of Bedburg-Königshoven (9,000–8,000 years ago) in the Rhineland (Germany).[19] Similar remains of dogs have been found from the numerous waterlogged Mesolithic sites in Denmark, where the period is known as the Maglemosian.

Perhaps the most famous of all European Mesolithic sites is Star Carr in Yorkshire, England. This was a hunting camp on the edge of a lake that was occupied periodically for a few centuries, more than 9,000 years ago, by groups of people who left everything they used and the remains of all the animals they killed to be preserved in the lakeshore mud. Their belongings, which were excavated in the 1950s, included wonderfully complete masks cut from the frontal bones of red deer stags with the antlers thinned down and with holes at the side for tying onto the hunter's face.[20] It is presumed that the masks were either worn by the hunters as decoys in the chase of red deer or perhaps they were also worn in ritual dances. Together with all the fascinating debris left by the hunters, there was the nearly complete skull of a large wolf, the partial skull of a young dog of about five months of age, and the left and right femurs and a tibia of another adult dog.[21]

More recently, in 1985, during excavation of the nearby waterlogged site of Seamer Carr, the neck vertebrae of a dog were retrieved, which match in size the skull of the subadult dog from Star Carr. These bones yielded stable carbon isotope ratios of -14.67 percent and -16.97 percent, which reveal that the dog had obtained a significant part of its food from a marine source.[22] Because the site of Seamer Carr, although a few hundred years later in date, was comparable with Star Carr and yielded very similar animal remains, it was at first postulated that the Mesolithic inhabitants who lived at the two sites also spent much of their year nearer the coast and subsisted on fishing. However, a recent carbon isotope analysis of bone samples from a number of species from Star Carr, including the dog and wolf, has shown that they all have much higher carbon isotope ratios with an average of -22 percent.[23] This shows that, although there were no human remains at Star Carr, it can be deduced that the hunters' diet at this rather earlier site was from terrestrial animals, and there were plenty of deer and other mammals and birds around the lake to support the human community. Over the next 100 years the wildlife may have become scarce (perhaps from overhunting), so the people had to move away, and Seamer Carr was only a temporary camp for hunters who spent more time at the coast as fishermen. Certainly, there have never again been finds in such prolific numbers and diversity of species at any other inland site as are represented from Star Carr. Besides the remains of aurochs (*Bos primigenius*), wild horse, and red deer, the large herbivores that were hunted included the latest dated remains in Britain of the elk (or moose, the largest of all deer, *Alces alces*).[24]

In the early Mesolithic of northern Europe there is no evidence for the cultivation of plants or for the domestication of any species of animal other than the dog. Britain was still joined in short stretches to the Continent, and there was still an abundance of large herbivores that could be hunted: aurochs, wild horse, red deer, and wild boar, while the shorelines could provide a munificence of seafood. Ireland, however, was cut off by the sea, and although hunter-gatherers had reached it and were living there, the only large mammals that could provide meat were red deer and wild boar, and it is possible that even these had been taken there by people in their boats.[25]

The economy of the inland-living Mesolithic people of Britain and much of Europe was based on the red deer (*Cervus elaphus*), which provided them with meat for food, hides for clothing and shelter, and bone and antler for tools and weapons. Farther north and up into the Arctic, human populations depended (and still depend) on the reindeer (*Rangifer tarandus*) for all their resources. Of all the manifold relationships that humans have had with hoofed animals, that with reindeer is unique, and it has also had the longest duration. From 20,000 years ago until the melting of the ice, reindeer were hunted as they lived and migrated over Europe, as far south as Spain. By the beginning of the Holocene, 10,000 years ago, the reindeer, which are supremely adapted in their physiology and diet to life in arctic conditions, had retreated to northern and arctic Europe and Asia.

The pastoralist and hunting peoples of the tundra and taiga (boreal forest) regions of Siberia and northern Russia have evolved many different ways of living with and off reindeer. The most ancient exploitation of reindeer is clearly by hunting the wild deer for their meat, hides, antlers, bones, and other bodily products.

The instinct for reindeer to migrate has made it possible for humans to share the arctic environment with them.[26] Over the millennia, until the present day, the attacks of biting insects have been the driving force for reindeer migrations over hundreds of miles, often followed by their only Holocene animal predator, the wolf. Unlike wolves, however, human hunters aim to intercept rather than follow moving herds of prey.[27] In their migrations between summer and winter feeding grounds, herds of reindeer follow exactly the same routes year after year over many generations. All that human hunters have to do is to position themselves at places where there will be bottlenecks during the migration, such as canyons or river crossings, where the reindeer can be easily speared or driven to their deaths. Many such sites have been identified in the archaeological record. Other ways of killing reindeer, as recorded for hunters in modern times, are with snares and by driving them into nets or pitfalls. Among those peoples who possess tame reindeer, live decoys are used to lure the wild prey near enough to shoot or spear.[28]

Reindeer are reputed to be among the easiest animals to tame. They are highly social, of gentle disposition and manageable size, and they are readily drawn to a lick of salt or human urine. For thousands of years they have been attracted to human settlements, not only for these nutrients but also to the smoke from household fires. This would help to keep off the millions of insect pests that plague them in the summer months, for each reindeer can lose up to 125 grams of blood a day from the bites of mosquitoes.[29]

Although individual reindeer probably lived as tamed animals since before the end of the last Ice Age, it seems that fully domesticated herds with individual reindeer that were used for draft and riding were not common until about 3,000 years ago. All the evidence comes from rock drawings in the Sayan Mountains on the border of Siberia and Mongolia, which combine the images of reindeer with other domestic animals: dogs, camels, and goats. It is believed that the reindeer sledge was copied from dog sledges, which had long been used in the far north.[30]

From the melting of the ice, 10,000 years ago, until the first evidence for the domestication of plants and animals in Europe, around 3,000 years had to elapse, during which time sea levels rose and forests spread over the landscape. The faunal composition changed with the

Figure 2. A caravan of Arctic reindeer herders on the move. (Reproduced with permission from Øyvind Ravna's Photo Collection.)

change of the ecosystems, and only the hunter-gatherers, as the master predators, remained the same, perfecting their strategies for social living and increasing in numbers. Through all the changes one thing that remained the same would have been the basic tenets of human behavior, and one of these tenets is the instinct to nurture (as discussed in the Introduction), expressed in the keeping of tame animals. However, material evidence for taming in the archaeological record is bound to be an extremely rare occurrence. Two examples, one of a cat and the second of a bear, are given below.

Apart from the Palaeolithic and Mesolithic finds of canid remains that can be identified as "dogs," the earliest evidence for a tamed animal in Europe is the skeleton of a cat with a human burial in the site of Shillourokambos in Cyprus (see figure 6). The site is dated to around 9,500 years ago, and as there is no fossil evidence for wild cats (*Felis silvestris*) on Cyprus, it is evident that the animal was brought to the island by human agency. It may be argued that wild cats were carried in boats to Cyprus without being tamed animals, but as this skeleton was found buried next to that of a human and in the same orientation, it can be assumed that the cat and the human had a special relationship.[31] This is the most ancient find of a cat in Cyprus, but it is not the first, for the mandible of a cat had been excavated from the site of Khirokitia in 1983, dated to around 6000 BCE.[32]

Proof that the Mesolithic hunter-gatherers of Switzerland kept a brown bear (*Ursus arctos*) tied up with a rope is evidenced by the finding of a lower jaw from the site of La Grande Rivoire in the northern Alps, dated to about 6,000 years ago. The bone is deformed in a way that shows that a thong had been tied around the jaw when the bear was a cub, and this had pushed the teeth out of position as they erupted. The thong must have been tied when the bear was about six months old, and it had lived until it was about six years old.[33]

Although finds that show evidence of taming in the early prehistoric period, such as this bear jaw, are rare finds indeed, evidence for the movement of live animals to new areas is easier

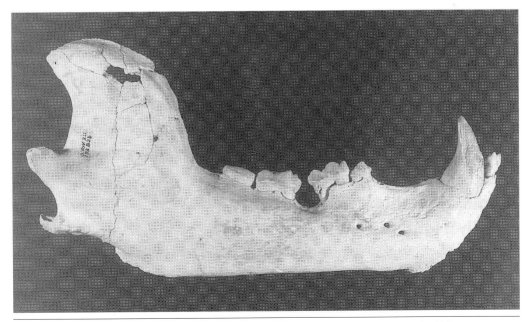

Figure 3. Right side of the lower jaw of a bear that had been tied with a "rope," dated to around 6,000 years ago. (Reproduced with permission from Louis Chaix.)

to deduce from excavation of their skeletal remains where there is no previous record of their fossil history. It was not only cats that were taken to Cyprus, for the bones of Mesopotamian fallow deer (*Dama mesopotamica*) were excavated from the sixth millennium BCE sites of Khirokitia and Erimi.[34] They must have been brought in boats from the Levant, where they are still to be found living wild in small numbers.

It is probable that people began to move dogs, cats, deer, and many other species of animals with them as they traveled over the continents and onto the islands of Europe and Asia at the end of the Ice Age, and this set the stage for the later, slow migration of domestic livestock species from their centers of origin.

Settlement and Domestication in Eurasia

As the ice sheets melted in northern Europe and Asia, the climate of the Mediterranean region and western Asia became warmer and wetter, and the nomadic people, like those in the north, flourished by hunting the local wildlife and gathering the abundant edible plants and wild cereals. The earliest period after the end of the Ice Age in this region is known as the Epipalaeolithic, and like the Mesolithic people of Europe, the hunters used arrows tipped with small sharp flakes, or microliths. The Epipalaeolithic has a different terminology in different regions of western Asia. In the Levant it is called the Kebaran, named after the type site of Kebara Cave, south of Haifa, dated from around 14,000 to 12,000 years ago.

The Kebaran period was a time of high productivity for humans, animals, and plants. Huge herds of gazelles must have inhabited the grasslands, and their bones predominate in the faunal assemblages from all the sites.[1] The later Kebaran period, known as the Geometric Kebaran after the shape of the microliths, overlapped with the period, also Mesolithic, that is known in the Levant as the Natufian. Kebaran sites, followed by the Natufian, are of the greatest importance for revealing the earliest evidence of communities living a settled existence with the beginnings of plant cultivation. It is here that can be seen, as James Mellaart wrote in his now classic book, *The Neolithic in the Near East*,

> the dimly emerging picture of an increasing awareness of the potentialities of the environment, the first steps towards food conservation through herding, wild grain collecting and its preparation as food and a trend, however limited, to sedentary life. Such features did not appear everywhere at the same time, they were as yet isolated and experimental, yet they constituted the embryo stage of plant and animal domestication which was ultimately to carry man from a hunting existence to that of the farmer and trader.[2]

Mellaart was correct in asserting that the Natufian was the crucial period for the beginnings of agriculture. However, much archaeological discovery has continued at the sites since Mellaart's book was written in 1975, and there has been much more discussion about what prompted the people to leave their way of life as nomadic hunters and settle down in semi-permanent villages, where finds of sickle blades are a witness to the harvesting of wild cereals.[3] Most of the meat for the communities was still obtained from hunting gazelles, but the faunal remains from Natufian sites show a greater diversity of species than at earlier sites, with birds, tortoises, snakes, and mollusks being represented.

The Late Natufian, dated to 10800–9500 BCE, occurred during the climatic period known as the Younger Dryas when the warm and wet woodland areas became colder and drier and the carrying capacity of the land was reduced. This has tempted archaeologists to propose the hypothesis that the Natufians, whose population numbers had greatly increased in the earlier period, were driven to live a semisedentary existence and expand their food resources by a shortage of wild animals to hunt.

Whatever prompted the Natufian people to change their way of life, like the Mesolithic people of Star Carr in England and Bedburg in Belgium, they owned "dogs" whose skeletal remains are distinctly different from those of wolves. The best known of these is the skeleton of a puppy from the site of Ein Mallaha.[4] The site, dated to 9600 BCE, is near Huleh Lake in the upper Jordan Valley. Its inhabitants were hunter-gatherers who, like other Natufians, were on the verge of becoming agriculturalists. They lived in round stone dwellings, used stone pestles and mortars for grinding cereals, and buried their dead in stone-covered tombs. In one of these tombs at the entrance of a dwelling, the skeleton of an elderly human was found together with the skeleton of a puppy of between four and five months of age. The human skeleton lay on its right side, in a flexed position, with a hand on the thorax of the puppy. The canid skeleton is too large to be a jackal and was either a tamed wolf or a dog.

It is more difficult in sites of this region than in Europe to distinguish between the skeletal remains of juvenile wolf and dog because the local Arabian wolf, *Canis lupus arabs*, is a very small subspecies, but there can be no doubt of the cultural affinity between human and canid in the close association of these two skeletons.[5] With adult canids, there is no difficulty in distinguishing the skulls of the earliest "dogs" from those of the Arabian wolf, and since the finding of the Ein Mallaha puppy skeleton, two fairly complete remains of "dogs" that were associated with three human skeletons were excavated from the site of Hayonim Terrace, dated to the Late Natufian.[6] The skulls of these "dogs" had shortened facial regions, and their carnassial teeth were reduced in size compared to those of a wolf, these being two of the characters that are used to distinguish dog from wolf.[7]

THE FIRST DOMESTIC CATS

It used to be generally believed that the ancient Egyptians were the first to keep cats as pets, some 3,600 years ago, but recent archaeological and genetic evidence has shown that their domestication has a much older history. Furthermore, the close relationship between cats and people first occurred not in Egypt but in the Near East. What has not been in discussion is the certain knowledge that all domestic cats (*Felis catus*) are descended from one species of felid, the wild cat, *Felis silvestris*, which has a wide distribution over much of Europe, Africa, and Asia. Molecular studies have shown that *Felis silvestris* has five genetic clusters or lineages that inhabit separate geographical areas: *F. silvestris silvestris* in Europe, *F. s. ornata* in Central Asia, *F. s. bieti* in China, *F. s. cafra* in southern Africa, and *F. s. lybica* in western Asia. All the hundreds of living domestic cats that were sampled belonged only to the *F. s. lybica* cluster, and from this it was deduced that there must have been a single original locality for the first group of domestic cats, and this has to be the Near East.[8] Probably wild cats began to move near human settlements soon after the beginnings of cereal cultivation and the storage of grain in the Pre-Pottery Neolithic B (PPNB) period. Stores of grain in heaps or in pits would have

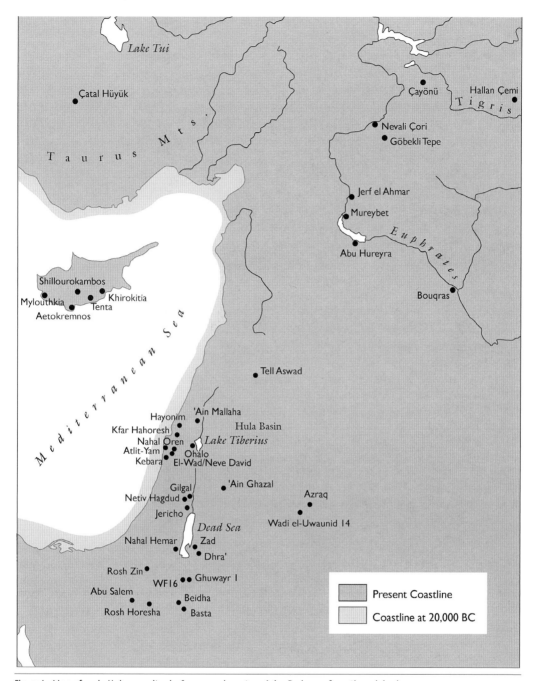

Figure 4. Map of early Holocene sites in Cyprus and western Asia. Redrawn from the original.

attracted rodents that would have been easy prey for the cats. One of these rodents was the house mouse (*Mus domesticus*), which had slowly moved west from northern India.[9]

The finding of the burial of a puppy with a human in the Natufian site of Ein Mallaha is paralleled by the burial of a young cat with a human at the early Neolithic site of Shillouro-kambos on the island Cyprus, which is dated to 9,500 years ago. As there is no fossil evidence

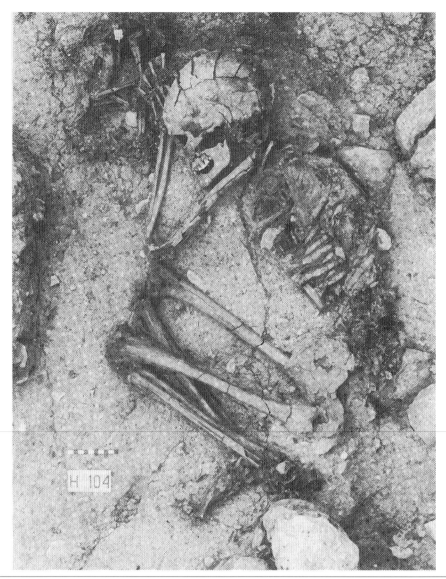

Figure 5. The skeleton of a puppy under the left hand of a human skeleton at the site of Ein Mallaha, northern Israel, c. 9600 BCE. (Source: S. J. M. Davis and F. R. Valla, "Evidence for Domestication of the Dog 12,000 Years Ago in the Natufian of Israel," *Nature* 276, no. 5688 [1978]: 608, Figure 1. Reproduced with permission from *Nature*.)

for there ever having been a species of wild felid living on Cyprus, this cat or its forebears must have been brought to the island, and its burial in close association with the human skeleton shows that the person and the cat had a special relationship in life.[10]

Cats do not fit the patterns of behavior of species that were "preadapted" for domestication—that is, species that are highly social and live in herds or packs in a dominance hierarchy to which the members are actively submissive. Cats are solitary hunters; they do not obey commands but "have their own agenda"; and they live in a more clearly commensal relationship

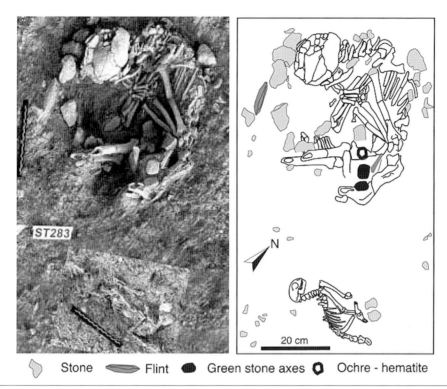

Stone — Flint ● Green stone axes ◯ Ochre - hematite

Figure 6. Photo (*left*) and plan (*right*) showing the burial of the cat (*lower skeleton*) and the human (*above*) at Shillouro-kambos, Cyprus, c. 9500 BCE. (Source: J.-D. Vigne, J. Guilaine, K. Debue, L. Haye, and P. Gérard, "Early Taming of the Cat in Cyprus," *Science* [Brevia] 304, no. 5668 [2004]: 259. Reproduced with permission from *Science*.)

with their human partners than do other domesticates. On the other hand, their bodily shape, especially that of kittens, intrinsically appeals to the human nurturing instinct, and domestic cats have made the most of this for thousands of years.[11]

THE BEGINNINGS OF LIVESTOCK HUSBANDRY

Gradually, over several thousand years, the Epipalaeolithic people of western Asia became more dependent on cultivated cereals and pulses for their staple diet, and the wild gazelles, cattle, goats, and boar that they hunted were replaced by domesticated livestock. Why this happened where and when it did, in the so-called Fertile Crescent, has puzzled archaeologists for nearly a century, but of course the findings partly follow the concentration of excavations in this area, and there is good evidence to indicate that there was equally early beginnings of plant cultivation in eastern Asia and the Americas.[12]

The result of the great number of sites excavated in western Asia has led to many careful analyses of the faunal remains and much discussion on the causes for the change from hunting to the keeping of tamed livestock as a "walking larder." This may be seen as the most important change in social and cultural behavior to have occurred throughout the history of the human species. It meant that, unlike animal predators, humans could control their food source and store their resources for the future, and it is likely that the impetus for this unique and crucially valuable ability arose from the overhunting of wild prey by a rapidly increasing human population around 10,000 years ago in western Asia, at the same time as changes to a drier climate affected the distribution of wild herbivore populations.

Although very many prehistoric sites have now been excavated in the Fertile Crescent, the Jericho Tell in the Jordan Valley still illustrates the archaeological evidence for the development of a farming economy from a hunting economy. The tell was excavated between 1952 and 1958 by Dame Kathleen Kenyon,[13] and the site has habitation levels that extend from the Natufian almost continuously through the Neolithic and Bronze Ages to the Byzantine. Following the Natufian, from around 10,000 years ago and for 1,000 years, in the period known as the Pre-Pottery Neolithic A (PPNA), the people of Jericho lived in a large village covering about 4 hectares. They were hunters, numbering perhaps 2,000 people, who lived in mud-brick houses with defensive walls of stone. They did not make pottery. Nearly all their meat still came from gazelles, but they also hunted wild cattle (*Bos primigenius*), wild boar, wild ass, goats, sheep, hares, birds, and other small animals, including foxes. It is possible that they cultivated some cereal crops.

Figure 7. Aerial view from the north of the Jericho Tell in the 1950s. (Source: K. M. Kenyon, Digging Up Jericho [London: Ernest Benn, 1957], 104, Plate 2B. Reproduced with permission from the Council for British Research in the Levant.)

THE CHANGE FROM HUNTING GAZELLES TO HERDING GOATS AND SHEEP

The next period at Jericho, as well as at many other Near Eastern sites, the PPNB, began around 9,000 years ago. The people still made no pottery, but they lived in houses of considerable size and built round courtyards, and the imprints of rush mats have been found on the floors. The cultivation of cereal crops became established in this period, and changes can be seen in the bones and horn cores of a small proportion of the goat and sheep remains that signify domestication.[14] Although the anatomical signs of domestication are few in number, there was a very large increase in numbers of goat remains and a decrease in gazelle remains, which is another sign that wild goats were being enfolded around the habitations of these first farmers, and hunted gazelles were a dwindling source of meat. In the PPNA levels of the tell, the number of gazelle bones and teeth constituted 55.4 percent of the total number of food animals, and goat bones and teeth constituted 4.3 percent, while in the PPNB levels the percentages were 50.2 percent for goats and 14.1 percent for gazelles.[15] Probably there were also attempts to herd gazelles, but this would not have been successful for gazelles are too flighty to be constrained, and they have a complex pattern of territorial behavior in the rutting season that makes it difficult to breed them in captivity.[16]

The ancestor of all domestic goats is the bezoar goat (*Capra aegagrus*), which inhabited the rocky slopes of mountains of the Levant in the early prehistoric period. Bones of this

Figure 8. Room in a PPNB house with plastered floors, walls, and doorways. (Source: K. M. Kenyon, *Digging Up Jericho* [London: Ernest Benn, 1957], 112, Plate 10A. Reproduced with permission from the Council for British Research in the Levant.)

wild goat have been identified from sites in Lebanon, Syria, Jordan, and Israel, as far south as Beidha (see figure 4), where its distribution overlapped with another species of wild goat, the ibex (*Capra ibex*), for which there is no genetic evidence for domestication.[17] By the late PPNB, around 8,000 years ago, goat remains were replacing gazelles and were spreading to sites farther north and east.

The ancestor of domestic sheep is the Asiatic mouflon (*Ovis orientalis*), whose natural habitat is on the foothills and slopes of hills and mountains in Turkey, northern Syria, northern Iraq, and Iran, at a lower altitude than that of the wild goats. The oldest site in western Asia from which bones of the wild mouflon have been identified is Zawi Chemi Shanidar in northeast Iraq dated to around 10,000 years ago.[18] Five thousand years later, the so-called Neolithic revolution was complete, and by the Early Bronze Age period of the Jericho Tell, domestic goats and sheep made up 75.8 percent of the animal remains while gazelles made up a mere 5.5 percent.

There are probably many complicated reasons why the early Holocene people of the Fertile Crescent gave up being hunter-gatherers and became herders of livestock, but the reason for why their flocks were of goats and sheep rather than gazelles is easier to understand and lies in the natural behavior patterns of the species. Gazelles live in open country and depend on speed for flight from predators. Like most species of deer and antelopes, gazelles live in herds, but their social system is not based on a dominance hierarchy, although they do defend a territory within the core area of their home range. Goats and sheep, because they are mountain animals without many predators, are not fine-tuned for escape. The herds follow a single dominant leader, and they live in a home range without defense of a territory. It is clear to see, therefore, how goats and sheep can adopt a human herder as their leader and the surrounding pasture as their home range.[19] Among pastoralists it is traditional, and probably has been for thousands of years, for the shepherd to guard and move his flocks with trained dogs and with a bellwether, a female or a castrated male goat or sheep that has been specially reared by the shepherd to be a natural flock leader.[20]

PIGS

In the early Holocene, wild boar (*Sus scrofa*) were distributed over most of the Northern Hemisphere, and their skeletal remains are found in almost all Epipalaeolithic and later hunter-gatherer sites in western Asia. At Jericho, the number of bones and teeth of wild boar gradually increased from 10.2 percent of the total remains of the primary food animals in the PPNA to 15.2 percent in the PPNB. In this latter period, pigs may have lived in a commensal relationship with the villagers, but there is little evidence of this in the pigs' skeletal remains, although the bones are somewhat smaller than those from the preceding period.[21]

Pigs are easy to tame when young, and, like dogs, they are useful scavengers that will clean up the detritus from village life. The earliest definite evidence for domestication of pigs comes from several sites in Anatolia, principally that of Çayönü Tepesi, which has habitation levels dating from 8,000 to 6,000 years ago.[22]

CATTLE

Throughout the Stone Age and early Neolithic periods, wild cattle (also called aurochs, *Bos primigenius*) were hunted for their meat, horns, and presumably also for their hides, sinews, and other useful products, although these have not been preserved in the archaeozoological record. The extinct aurochs had much the same distribution over the Northern Hemisphere as wild boar, and the huge bulls, which stood over 6 feet at the shoulder, must have been even more formidable as prey.[23] The remains of hunted aurochs made up 11.7 percent of the total number of animal remains from the PPNB levels of Jericho, but by the Early Bronze Age this had fallen to 0.2 percent. After these 5,000 years of exploitation, wild cattle had probably become very scarce, and they were being replaced by the domestic form, which made up 12 percent of the total number of animal remains.[24] By this period domestic cattle were being used for draft as well as for providing milk and meat.

Jericho was not, however, a center for cattle domestication or for the earliest evidence for the worship of bulls, which are sited at Çatal Hüyük in the alluvial Konya plain of Anatolia (see figure 4). This vast Neolithic mound, covering 32 acres, was excavated first by James Mellaart between 1961 and 1965,[25] and since 1993 by an international team of archaeologists led by Ian Hodder.[26] The earliest levels at Çatal Hüyük are dated to 6500 BCE (the late PPNB), and the site was inhabited as a town with settled agriculture and domestic animals for the next 2,000 years. The livestock economy appears to have been based on cattle rather than goats and sheep as in the Levant. Mellaart calculated that there could have been 1,000 houses in the town in its heyday, and the average population was probably 5,000–6,000 inhabitants.[27]

Figure 9. Aurochs horn cores set in clay bucrania. (Owner of image: Çatal Hüyük. Source: Flickr.com.)

Çatal Hüyük is unique in its early period for the wall paintings of bulls and for the horn cores of aurochs that were discovered in the excavated shrines.

Mellaart described the shrines in these terms:

> The numerous shrines at Çatal Hüyük take us a step further, for here there are scenes depicted in plaster reliefs that defy materialistic interpretation. Women do not give birth to bulls' or rams' heads, and certainly not on top of a doorway from which peer the ferocious heads of three super-imposed bulls, made the more lively by the use of actual horn cores. Nor is it very usual to find 6-foot-tall Siamese female twins, one of which gives birth to a huge bull's head, on top of which there appears another, smaller one. The benches found in several shrines, in which one, three, five or seven pairs of horn cores of aurochs are embedded, are matched in simpler form by the bucrania, brick structures with a pair of horns, set on the edge of platforms, found in others.[28]

Mellaart believed that the settlement at Çatal Hüyük represented a local Anatolian tradition of agriculture and ritual, but more recent excavations have shown that the aurochs was revered as a sacrificial animal during the PPNB throughout the Near East. One example has been excavated in the "*Bos* pit" at the mortuary site of Kfar HaHoresh in northern Israel.[29] The food animals identified from this PPNB site comprise the same species as at other sites of the period. Gazelles were the most commonly hunted animals, followed by wild goats, with aurochs, wild boar, and deer being represented in lower numbers; red fox, reptiles, rodents, fish, and birds were also present. More than sixty humans were buried at the site, but one burial was unique in having the skeletal parts of eight aurochs buried at the bottom of a pit with the human skeleton above the cattle bones. Many of the bones were still in articulation, which showed that they were the remains of joints that must have provided a minimum of 500 kilograms of meat. The excavators, Nigel Goring-Morris and Liora Kolska Horwitz, have inferred that the "*Bos* pit" is perhaps the earliest evidence for a mortuary feast in the Near East. Whether the aurochs that were killed for this feast were culturally managed and were the direct ancestors of the fully domestic cattle of later times cannot be assessed without a detailed genetic analysis of their bones. However, there is no doubt that the aurochs played a special role in settlements of the PPNB period, which stretched from Central Anatolia to the Levant, and that by 8,500 years ago this giant wild ox was beginning to be tamed.

THE FIRST DOMESTIC DONKEYS IN WESTERN ASIA

The earliest domestic goats and sheep would probably have been killed for their meat in the manner that had been traditional with wild animals for thousands of years. There must then have been a slow progression to using goat and sheep fleeces for spinning and making cloth as well as the innovation of taking their milk. This was followed with the domestication of cattle for their meat and milk, and toward the end of the fourth millennium BCE by the invention of wooden sledges with runners, and some with wooden disk wheels, that were harnessed to castrated or docile animals and used for traction. The earliest evidence for these primitive carts was found as pictographs inscribed on clay tablets at the site of Uruk in southern Mesopotamia, dated to around 3200–3100 BCE.[30]

Perhaps as much as 1,000 years later tamed asses also began to be harnessed, and their

Figure 10. Pictographs of sledges on clay tablets from the site of Uruk, Mesopotamia. 1. Simple sledges and two with disk wheels; 2. Sledge harnessed to an ox and with a driver. (Source: M. A. Littauer and J. H. Crouwel, *Wheeled Vehicles and Ridden Animals in the Ancient Near East* [Leiden/Köln: E. J. Brill, 1979], Figures 1 and 2. Reproduced with permission from J. H. Crouwel.)

remains are found from Sumerian sites of the third millennium BCE.[31] In Mesopotamia, there are numerous finds of incised images of wheeled vehicles as well as metal and terra-cotta models and burials of whole skeletons of equids. These equids are all asses, which in the early days of excavation of Near Eastern sites were identified as tamed onagers, the wild Asian ass (*Equus hemionus*).[32] The reason for this was that onagers were the common wild asses of western Asia, so it was logical to assume that they would have been domesticated. However, if the onager had been used as a domestic ass throughout western Asia, it would be expected that the modern domestic ass of the region would be descended from this ancient stock, but this is not so. As far as is known, all the donkeys of Asia (and indeed throughout the world) are descended from the African wild ass (*Equus africanus*), and if they interbreed with onagers, their offspring are sterile. Furthermore, osteological examination of the remains of asses from the Sumerian sites has shown that the bones and teeth can nearly all be identified to the *E. africanus* morphology and not *E. hemionus.*

Archaeologists have always assumed that the donkey (*Equus asinus*) originated from Egypt with the domestication of the indigenous wild ass (*Equus africanus*), but, as discussed by Hans-Peter Uerpmann, there are equivocal equid remains from a few early sites in western Asia that have been identified as wild *E. africanus.*[33] The importance of these bones is that they provide a basis for the presumption that the first domestication of the donkey took place from the wild species in western Asia and not in Egypt. However, recent molecular analysis of all the living species of asses and onagers clearly shows that all domestic donkeys are descended from the two subspecies of African wild asses (*E. africanus africanus* and *E. africanus somaliensis*). The authors of the genetic study further hypothesize that there were therefore two centers of domestication of the donkey and that both were in northeast Africa.[34]

A few of the equid skeletal remains from Mesopotamia, however, do appear to be hybrids between donkey and onager, and this is supported by the terms, written in cuneiform on clay tablets, that the Sumerians used for their different equids. These can be summarized and translated after the work of Nicholas Postgate:[35]

anse = generic term for any equid or domestic ass (*E. asinus*)
anse-DUN or anse-LIBIR = domestic ass (*E. asinus*)
anse-eden-na = onager (*E. hemionus*)
anse-BARxAN = hybrid *E. asinus* x *E. hemionus*

Where skeletons of equids were in pairs as discovered at the site of a Sumerian burial at Abu Salabikh, Postgate believes that they must have been yoked for traction and were possibly drawing a small chariot, which has disappeared with erosion in the tomb.[36]

By around the middle of the third millennium BCE, chariots with wooden wheels were in use for warfare and were drawn by donkeys or by hybrid donkeys x onagers. The most famous representation of chariots of this date is on the wooden box known as the Standard of Ur. The box (which has been reconstructed and is in the British Museum) has two panels showing a battle scene and a peace scene. The "chariots" are in fact four-wheeled wagons that would have been very unstable when driven at any speed, and they were soon replaced by two-wheeled chariots throughout the land of Sumer.

Toward the end of the third millennium BCE, the use of donkeys as pack animals and for traction as well as possibly for riding was widespread throughout western Asia, although their numbers were probably still relatively small, and they may only have been owned by the elite. Donkeys also had a ritual function, and their skeletons have been found where they have clearly been buried as sacrificial offerings. At the Akkadian site of Tell Brak in Syria (c. 2200 BCE), the complete skeletons of a dog and a donkey were excavated and had clearly been buried as sacrifices, while close by, in the area of a small temple, another five skeletons of donkeys were found. An indication that the donkeys had been kept in stalls and also had been loaded and possibly ridden was provided by examination of the teeth and vertebrae. The incisor teeth of two of the donkeys were notched in a way that is commonly found with equids that have been chewing on wood and is known as crib-biting. From this it can be inferred that the donkeys had been confined and had been chewing on the wood of their stall. In another of the skeletons (an aged male) the first upper cheek tooth was unnaturally worn in a manner that is characteristic of equids that chew on a bit, and in yet another (an aged female) the neural spines of the thoracic vertebrae had outgrowths that probably resulted from heavy loading of the donkey's back.[37]

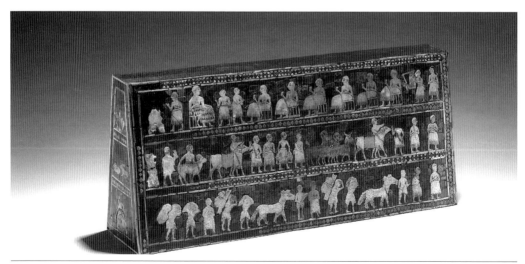

Figure 11. The war panel from the Standard of Ur. (Copyright Trustees of the British Museum, reproduced with permission.)

THE FIRST DOMESTIC HORSES IN WESTERN ASIA

For much of the twentieth century it was accepted that domestication of the horse first took place in Central Asia, based on the distribution of the relative abundance of fossil remains of wild horses together with the bones and teeth of horses from early prehistoric sites that appeared from archaeozoological analysis to be domestic, or at least to have been under some human control. However, the "where and when" of the earliest domestic horses is a subject that has been more difficult to disentangle than perhaps for any other species of ungulate. It is only recently that molecular studies have shown that more than seventy-seven maternal lineages have been recruited from the wild in the domestication process.[38] This implies multiple centers of breeding from wild mares, but it could only occur where remains of wild horses have been found in large numbers, and the Near East is not one of those regions.

Toward the end of the last Ice Age wild horses (*Equus ferus*) roamed over much of Eurasia, but with the warming of the climate and the spread of forests, and probably also as a result of extensive hunting by humans, the herds became scarce and died out in many regions, including the Levant, where Pleistocene fossils of horses have been found. When remains of horses are then recovered from archaeological sites with a chronological gap of more than 7,000 years, it may be safely assumed that they came from introduced domesticated animals. This is the situation in the Levant, where Caroline Grigson has identified bones that can only be from domesticated horses from eight Chalcolithic sites in the northern Negev (Israel).[39] The dating of these sites to the fourth millennium BCE was surprising, as the general consensus among archaeozoologists has been that domestic horses were introduced to the Near East in the Early Bronze Age in the first half of the third millennium BCE.[40]

Joan Oates, who with her late husband, David Oates, was the renowned excavator of the vast Akkadian site of Tell Brak in Syria, has reported that

> the word for horse first appears in the Ur III period, conventionally dated to the last century of the third millennium BCE. In the Sumerian anse.zi.zi (later anse.kur.ra, the "ass of the mountains") derived from akkadian sisu (sisa'u), itself probably an Indo-European loan word. That this must refer to domestic horses is clear from the contexts. At this time horses are always listed in small numbers, under the care of specific persons.[41]

The riding of equids is shown on a large number of clay figurines from the late third millennium BCE, but it is difficult to discern whether donkey, hybrid, or horse is depicted. One clay tablet from Ur, dated to 2037–2029 BCE, shows a rider on what is clearly a horse from its hanging mane and full tail. From the depictions of riders seated on the rump of the horse, it may be presumed the riding of donkeys was a commonplace activity before it was transferred to horses, for whereas the normal position for riding astride a donkey is far to the back in what is known as the "donkey seat," this would be a very uncomfortable and unstable position for riding a horse at speed. An even older transference can be traced from the use of reins and a nose ring for driving horses, which must have been derived from the use of cattle for draft.

Figure 12. A horse and rider on a clay plaque from Ur, 2037–2039 BCE. (Source: Joan Oates, "A Note on the Early Evidence for Horse and the Riding of Equids in Western Asia," in *Prehistoric Steppe Adaptation and the Horse*, ed. by Marsha Levine, Colin Renfrew, and Katie Boyle [Cambridge: McDonald Institute for Archaeological Research, 2003], 118, Figure 9.5. Reproduced with permission from McDonald Institute Publications.)

Figure 13. A horse and rider in the "donkey seat" on an early second millennium BCE clay plaque. (Source: M. A. Littauer and J. H. Crouwel, *Wheeled Vehicles and Ridden Animals in the Ancient Near East* [Leiden/Köln: E. J. Brill, 1979], Figure 37. Reproduced with permission from J. H. Crouwel.)

Figure 14. Detail of a seal impression from Kültepe (1950–1850 BCE) of equids harnessed to a chariot with reins and nose rings. (Source: M. A. Littauer and J. H. Crouwel, *Wheeled Vehicles and Ridden Animals in the Ancient Near East* [Leiden/Köln: E. J. Brill, 1979], Figure 29. Reproduced with permission from J. H. Crouwel.)

Arrival of Domesticates in Europe

EVER SINCE GORDON CHILDE, ONE OF THE MOST ERUDITE AND RENOWNED PREHIS-
torians of the twentieth century, wrote his classic work, *The Dawn of European Civilization*,
which was first published in 1925, there have been countless books, articles, and reviews
written on the spread of farming from the Fertile Crescent westward and north across the
European continent.[1] As Childe noted in 1958, in his preface to a later, more popular work,
the subject "is often buried under a forbidding accumulation of outlandish culture-names
and references to obscure periodicals."[2] Today, as might be expected, this accumulation has
grown enormously, but the core fact remains that the earliest knowledge about the cultivation
of crops and the husbanding of livestock came to Europe from southwest Asia. Whether this
knowledge came with the movement of people or by trade in plants and animals from one
region to another has been much argued, but the overall picture must be a complex pattern
of social and cultural change in which hunter-gatherer communities were transformed from
collectors to producers of their food and other essential resources. This transformation should
be seen as slow and affected by many factors both within the human communities and in the
environment around them.[3]

Although the last Palaeolithic hunters would no doubt have been skilled managers of their
prey, the material evidence for this is sparse, while the evidence for overhunting in the progressive scarcity over time of the bones of wild ungulates in Mesolithic sites is more evident. As
suggested in Chapter 1, the invention of the long-distance projectile in the form of bows and
arrows, in combination with domestic dogs, would have greatly increased the hunters' efficiency.

Following the first stage of close management in which ungulates would be kept alive as
a walking larder rather than killed in the hunt, the way of life of the last hunters/first farmers would have undergone significant changes, which in some communities would have been
rapidly accepted and in others would have been resisted. Animal dung would be used in the
cultivation of crops, for fuel, and for building; hair, wool, and hides would be used for clothing; milk would become a vital part of the diet; and cattle would provide traction. With the
alteration in the material and spiritual outlook on the world, religion would change.

The archaeozoological evidence for the so-called wave of advance for the spread of agriculture across Europe began in the eastern Mediterranean around 8,000 years ago, and it is
not surprising that it took 3,000 years to reach Britain and Scandinavia in the north, and the
Iberian Peninsula in the extreme west. Using radiocarbon dates that have been obtained for
relevant archaeological sites, the advance has been plotted by Julian Thomas in his review of
the Neolithic revolution as seen in figure 15, which was redrawn from the earlier model of
Ammerman and Cavalli-Sforza.[4]

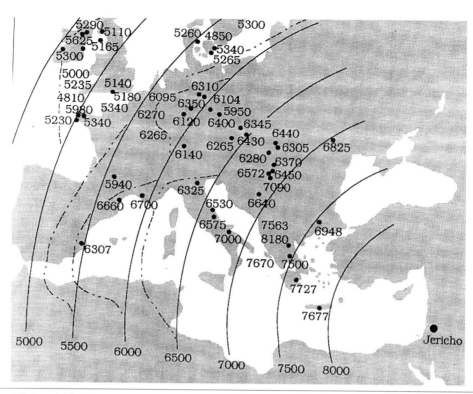

Figure 15. A model for the "wave of advance" of agriculture from the Near East across Europe, using radiocarbon dates in years before present. The broken lines represent probable regional variations in the rate of spread. (Source: Julian Thomas, "The Cultural Context of the First Use of Domesticates in Continental Central and Northwest Europe," in *The Origins and Spread of Agriculture and Pastoralism in Eurasia*, ed. by David R. Harris [London: UCL Press, 1996], 311, Figure 17.1. Reproduced with permission from Julian Thomas.)

The very beginning of the westward transport of livestock and other animals from the Near East appears to have begun on Cyprus, at the extraordinarily early date of at least 9,000 years ago. At the site of Shillourokambos there is osteological evidence that wild cattle (*Bos primigenius*), goats (*Capra aegagrus*), sheep (*Ovis orientalis*), Mesopotamian fallow deer (*Dama mesopotamica*), foxes (*Vulpes vulpes*), and cats (*Felis silvestris/catus*) had been carried in boats across the sea from the Anatolian or Levantine coasts. There are also two sizes of pig remains, corresponding to the wild (*Sus scrofa*) and the domestic (*Sus domesticus*). As there are no fossil remains of these species from earlier periods on Cyprus, it can be assumed from these finds that the animals were taken there by humans. Furthermore, the large size of the ungulate bones and teeth indicates that the animals were in an intermediate stage between the wild and the domestic. They were presumably released in the new environment, became feral, and were then hunted as wild animals.[5] There are many other early Neolithic sites on Cyprus apart from Shillourokambos, the most notable being the rather later site of Khirokitia from which the remains of sheep, goats, pigs, Mesopotamian fallow deer, and a small cat were first described by Judith King in 1953.[6]

A plausible scenario may be that there were still pigmy hippos living on the island of Cyprus when the first traveling hunters arrived there, perhaps 10,000 years ago. The overlap of humans with living hippos receives unconfirmed support from the finding of presumed knife cuts on fossil hippo bones.[7] The hippos would have been quickly exterminated, leaving the expanding human population without meat, for which the only remedy was to bring live ungulates from the mainland and let them live and breed freely on the island. This would have been possible if the animals were calves, weaned but still small enough to be carried onto boats, even if they were caught from wild stock.

The Mesopotamian fallow deer died out on Cyprus in historic times, as did the other imported wild species, apart from the sheep, which are still to be found living as a feral population in the mountains. These sheep, known as Cypriot mouflon (*Ovis orientalis ophion*), are genetically distinct from the local domestic breeds, and they are assumed to be descendants of the original wild sheep (*Ovis orientalis*) that were taken to Cyprus in prehistoric times with some introgression from later breeding with domestic sheep.[8]

Sheep that are relics of early prehistoric transport are also found on Corsica and Sardinia, where they had long been presumed to be an endemic wild species called the European mouflon (*Ovis musimon*), and they were believed to be ancestral to European domestic sheep. However, in the 1960s the mouflon came under study in the new science of archaeozoology, and it was realized that the only wild caprine (sheep or goat) anywhere in Europe at the end of the Pleistocene was the alpine ibex (*Capra ibex*) and that the mouflon, although it had every appearance of a wild sheep, had to have been taken to the islands of Corsica and Sardinia by human travelers in the same manner as the mouflon on Cyprus.[9]

On Crete and on the island of Montecristo, and on a few other small Mediterranean islands, there are wild goats that are also relics of ancient importation. The Cretan goats are known as agrimi (*Capra aegagrus cretica*); they have been studied by Liora Kolska Horwitz and Gila Kahila Bar-Gal, who state:

> The earliest, securely dated goat bones from Crete derive from . . . Knossos dating to the end of the 7th millennium B.C. . . . They were found together with remains of other domesticates—sheep, cattle, pig and dog. Goats in this sample and in the overlying Early Neolithic Ia were fewer in number than sheep, but considerably larger in size. . . . The size of sheep was comparable to that of contemporaneous domestic sheep populations from the Greek mainland and other domestic sheep samples from Europe. . . . Bones of large sized goats (identified as agrimi) are found together with those of domestic sheep and goat in various excavations on Crete dating from the 6th millennium B.C. onwards. In the Minoan period . . . the agrimi is clearly depicted in art works as a game animal. These data indicate that the agrimi was well established as a hunted species following its introduction in the Neolithic.[10]

In the intermediate shape of their horns, general morphology, and genetics, the feral sheep and goats of the Mediterranean islands provide clear proof of the human transport of living animals at the very beginning of the transition from hunting to farming, but, although their survival over many thousands of years is remarkable, they are not alone. A huge variety of species of wild and domestic animals have been taken to islands and indeed to new continents ever since that ancient time, and of course are still being moved around in their millions at the present day. It is not possible to assess the period in the past when a living, domestic species is likely to have reached a new location unless it has remained reproductively isolated from other breeds and has the same phenotypic morphology as the skeletal remains of the species excavated from a distinct archaeological period. This situation is only likely to occur on islands because in all other places there will have been outbreeding with new stock over hundreds or thousands of years. For this reason, the populations of feral sheep and goats on the Mediterranean islands hold such archaeozoological value. Another far-distant population of island sheep of equal importance is the Soay sheep of the islands of St. Kilda in the Outer Hebrides of Scotland. These small, brown, domestic sheep were most probably taken to the islands by Bronze Age people around 3,500 years ago. They have remained almost unchanged in morphology ever since, with their bones closely resembling the skeletal remains of sheep from Bronze Age sites on the British mainland.[11]

Beginning in Greece in the Early Neolithic, as described by Paul Halstead, an integrated "package" of intensive horticulture and sheep slowly spread across southern Europe. In the later Neolithic the predominance of sheep is found to have given way to a more balanced assemblage of sheep, pigs, cattle, and goats at open archaeological sites and to a dominance of sheep and goats at cave sites. As caves are usually in hills and rocky landscapes, the dominance of goats can be seen as an adaptation to rough terrain and sparse browse. Halstead reports that over the course of the Neolithic in mainland Greece, there is a clear decrease in size of domestic pigs and cattle, with a widening in the difference between the remains of these domesticates and wild cattle (aurochs, *Bos primigenius*) and wild boar.[12]

GENETIC STUDIES ON THE ORIGINS AND DIFFUSION OF DOMESTICATES IN EUROPE AND WESTERN ASIA

Cattle

As the aurochs ranged widely over the whole of Europe throughout the prehistoric period, it has often been hypothesized that there were many local centers of domestication of cattle from the wild species. However, the detailed molecular studies of Dan Bradley and colleagues now suggest otherwise. First, the variation in microsatellite and mtDNA data is greater in domestic cattle in Anatolia and the Near East than in European populations, which implies that European variation is a subset of that found in the Near East. Second, a sample of *Bos primigenius* from Britain has mtDNA variation that is markedly different from that of modern cattle. Thus there does not appear to be detectable maternal contribution to the gene pool of European domestic cattle from locally domesticated wild cattle, but this does not preclude hybridization with wild European aurochs bulls.[13]

Sheep

The genetic origins of the domestic sheep are not so straightforward as for cattle, but all the indications from the molecular studies are that the Asian mouflon (*Ovis orientalis*) was the single ancestor of all breeds, including the European mouflon (*Ovis musimon*). The Asian species of wild sheep, the urial (*Ovis vignei*) and the argali (*Ovis ammon*), do not appear to have been involved in the ancestry of any modern breeds. Michael Bruford and Saffron Townsend, in their genetic study of modern breeds, found three domestic clades, which they suggest are derived from different *Ovis orientalis* subspecies or divergent populations, implying separate waves of migration with the first farmers from western Asia.[14]

Goats

As with sheep, no definite answers on the small-scale genetic origins or geographic locations of the origins of domestic goats have come from molecular analyses. However, as consistent with the evidence from archaeological sites in the Near East, the bezoar goat (*Capra aegagrus*) does appear to be the most probable ancestor of all domestic goats. The geneticist Gordon Luikart and colleagues have found that three genetic origins of goats are supported by the mtDNA and Y chromosome data. This fits with the three lineages found by Bruford and Townsend for sheep and could indicate that the first domestic sheep and goats traveled away from their centers of origin together.

According to Luikart and colleagues, the combined data suggest that there were at least three major centers of domestication for sheep and goats, but they point out that more recent hybridization with wild males, rather than ancient domestication, could explain the origin of a lineage, not only in goats and sheep but in all domestic ungulates.[15] These authors also computed the percentage of the total mtDNA variation in goats that was due to differences among continental populations and found that it was only about 10 percent, which is far lower than the variation in cattle between continents. Surprisingly, the variation between continents for humans is also about 10 percent, and the suggestion is that the relatively similar demographic history of goats and humans (compared to most vertebrate species) makes comparisons of their genetic structure "interesting and not necessarily inappropriate." The fact that the generation time in goats is four to six times shorter than in humans is used in this argument to explain that humans began their travels round the globe 50,000 to 100,000 years ago, while goats only began their association with humans 10,000 years ago, which makes the number of generations since population expansion in humans and goats fairly similar.[16]

Pigs

Unlike the aurochs, progenitor of domestic cattle, which was exterminated by human hunting and loss of habitat, wild boar (*Sus scrofa*), progenitor of domestic pigs, are still found in many parts of Eurasia, except in the coldest and driest regions, and they have ranged over all of Europe and most of Asia from the end of the last Ice Age. Archaeozoological evidence indicates that remains of the earliest domesticated pigs are to be found at sites in Anatolia, notably Çayönü Tepesi, where there is a 2,000-year sequence of animal remains dating from the ninth to the seventh millennia before present. In their mtDNA analyses Greger Larson and colleagues found that the genetic constitution (haplotypes) of wild boar from Turkey, Iran, and Armenia was unrelated to all the pigs examined from Europe, with the exception of one feral pig from Corsica, which clusters within the Turkish group. They have concluded that wild boar were independently domesticated in Europe, but that the Corsican pig may represent one of the last pigs in Europe whose ancestors were transported from Turkey during the Early Neolithic "wave of advance." From this, they suggest that the domestication of European wild boar may have resulted from the transfer of ideas, rather than of the actual animals.[17]

That pigs in Europe in the ancient world were not treated as mere scavengers but could be given care and special housing is described as early as the eighth century BCE in Homer's *Odyssey*.[18] Odysseus, on return from his travels, meets the faithful swineherd who had constructed a stone stockade inside which he had built twelve sties in each of which fifty sows slept on the ground and had their litters, while the boars slept outside.[19]

Horses

As discussed in Chapter 2, the origins of horse domestication have proved very difficult to disentangle, with the molecular evidence providing two competing theories. First, a restricted origin was postulated—that is, selective breeding occurred of a few tamed founders. The second and now generally accepted theory is that a large number of founders were domesticated over the wide Eurasian range of the wild horse.[20]

After the end of the Ice Age, wild horses slowly decreased in numbers and became localized in much of Europe and Asia, although they did not become finally extinct. However, Holocene remains of horses are only found in large numbers from prehistoric sites on the Eurasian steppes, notably in the Ukraine and Kazakhstan at excavations of Dereivka and sites of the Botai culture.[21] There has been much argument in the literature about whether the horses from these sites were wild and were killed for their meat or whether at least some of them were domesticated and were even ridden at a date of around 6,000 years ago.[22] Whatever the solution to this debate, it does seem probable from the molecular and archaeozoological evidence that Eurasian wild horses (*Equus ferus*) were domesticated at some period before the Bronze Age. These horses were more valuable than other livestock animals in their use for draft and riding, and their numbers rapidly increased, with a large number of mares being recruited from the wild across Europe and Asia to breed with and augment the domestic stock.[23]

A novel method of tracing the spread of domestic horses through time has been investigated

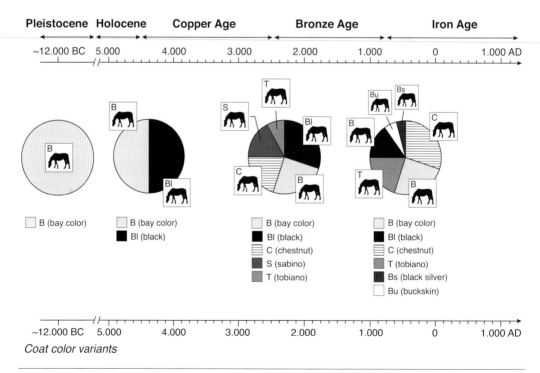

Figure 16. Timeline showing the rapid increase in coat color variation in horses during the Bronze Age. The size of sections in the pie charts corresponds to the frequency of the respective phenotype. Sample sizes from left to right are 16, 18, 26, and 29 specimens. (Source: Arne Ludwig et al., "Coat Color Variation at the Beginning of Horse Domestication," *Science* 324, no. 24 [April 2009]: 485. Redrawn with permission from *Science*.)

by Arne Ludvig and colleagues by means of analysis of the genes for coat color obtained from ancient samples of bone. The geneticists managed to deduce that wild Pleistocene horses had a single bay coat color, but that this rapidly became much more variable with domestication and selective breeding, so that by 2000 BCE (the Bronze Age) there were horses from Siberia and eastern Europe with a large number of different coat colors, as shown in figure 16. The earliest chestnut allele (MC1R gene) was identified in a Romanian sample from the late seventh millennium BCE, and the prevalence of chestnut horses increased rapidly, reaching 28 percent during the Bronze Age.

THE NORTHWARD SPREAD OF LIVESTOCK IN THE NEOLITHIC

The archaeological and biological evidence supports the theory that there was a slow spread of small-scale plant cultivation and livestock herding, termed agro-pastoralism, with an increasing immigration of Neolithic farmers west and north around the Mediterranean and into central and northwestern Europe. David Harris has summarized the approximate dating of this early phase of migration as follows: In southern Italy and the Dalmatian coast of the Adriatic, the earliest agricultural settlements are dated to the first half of the sixth millennium BCE with expansion across central and into northwest Europe in the sixth and fifth millennia BCE, where the archaeological typology is known as the Linearbandkeramik (LBK). In the Iberian Peninsula and southern France, the earliest Neolithic is not found before c. 4800 BCE. The second phase of expansion occurred after 4000–3500 BCE when Mesolithic groups living beyond the agricultural frontier around the coast of northern Portugal, Cantabrian Spain, western France, Belgium, the Netherlands, and Great Britain finally began to replace their hunter-gatherer way of life with farming. There is general agreement that the LBK represents colonization by immigrant farming groups who introduced goats, sheep, and cattle, as well as cultivated crops, to central Europe from southwestern Asia.[24]

THE BENEFITS OF FARMING

Ten thousand years ago the human populations of the world lived by hunting wild animals and gathering wild plants for food, while 5,000 years later the inhabitants of the most densely populated areas had become farmers. What were the motives for this radical change from the relatively easy way of life of the hunter-gatherer to the drudgery of managing livestock and sowing, protecting, and harvesting crops that had very low yields? In Europe, the driving force may have been the increasing human population, as first postulated by Mark Cohen.[25] And it looks increasingly likely that the answer was *milk*, as discussed below, but the change from hunting to farming was not a "revolution" in the usual meaning of "the overthrow of a system"; it stretched over 2,000 years and even then was not everywhere complete. The "Neolithic revolution" was the result of complex, interwoven social, cultural, and economic processes in the different regions of Eurasia that led to different traditions of farming, which have survived to the present day.

Milk

The "Neolithic revolution" used to be considered as a gradual progression for the provision of meat from the killing of animals by hunters to the killing of animals that were managed and bred in captivity—that is, they were kept as a walking larder. The next stage of this progress, as conceived by the late Andrew Sherratt, was in what he described as the "secondary products revolution."[26] This was presumed to have begun during the fourth millennium BCE in the Near East and involved keeping domestic livestock alive for their secondary products: milk, wool, traction, and riding. The archaeological evidence appeared to indicate that 1,000 years later, this all-important cultural change had spread to Europe and the rest of Asia.

Within the past few years, new biochemical techniques have led to the finding of organic residues on pottery sherds, which have proved that the pots had contained milk, and dating of the sherds has provided evidence that cattle were being milked much earlier than previously believed. As described by Richard Evershed and colleagues, the domestication of cattle, sheep, and goats had already taken place in the Near East by the eighth millennium BCE, but the first clear evidence for the use of traction and wool appears much later, from the late fifth and fourth millennia BCE. The organic residues on pottery sherds have provided direct evidence for the use of milk in northwestern Anatolia by the seventh millennium BCE, in the sixth millennium BCE in eastern Europe, and in the fourth millennium BCE in Britain.[27] This very early evidence for the milking of cattle puts a new slant on the whole perception of the transition from Mesolithic hunters to Neolithic farmers. It suggests that in areas where cattle were milked at a very early period, there was a basic need for increased nutrition (perhaps as a result of overhunting and climate change), and it indicates (what can already be presumed) that there was widespread and intensive knowledge of the natural behavior of animals and their management. It involves considerable skill and empathy to get a cow of a primitive breed to let down her milk, with the first essential being that the calf has to be present, as described by Amoroso and Jewell.[28]

Furthermore, it was not only cattle, and presumably goats and sheep, that were milked at the very beginning of the "Neolithic revolution," for there is also similar evidence from organic residues on pottery sherds that mares were milked at sites of the Botai culture in Kazakhstan, Central Asia, at around 3500 BCE. This confirms what has been much disputed, that the Botai horses were domesticated and were not wild horses killed for their meat.[29]

The very early use of fresh milk by the Neolithic people of central and northern Europe is supported by the molecular research of Yuval Itan and colleagues, who have identified a gene for the persistence of lactase in Europeans, while, with the exception of some African, Middle Eastern, and southern Asian peoples, it is absent, and described as lactose malabsorption, in most adults in the rest of the world. Lactase is the enzyme that in all infant mammals allows them to digest their mother's milk, and it normally disappears from the gut on weaning. It is unlikely that persistence of lactase would provide a selective advantage without a supply of fresh milk, and therefore the allele would only be inherited in societies that practice dairying and the drinking of fresh milk. Itan and colleagues believe that the allele for lactase persistence first underwent positive selection among dairying farmers around 7,500 years ago in a region between the central Balkans and central Europe, possibly in association with the dissemination of the Neolithic Linearbandkeramik.[30]

THE SPREAD INTO EUROPE OF THE FIRST WHEELED VEHICLES

Molecular analysis, as well as all the other scientific methods that can now be applied to archaeozoological specimens, can reveal an amazing amount of information about the history of domestication. However, excavation of material objects as well as organic remains will always provide the basis for learning about the history of technology and how domesticates were used in the past. One of the most important of human inventions was the wheeled vehicle, which made its first appearance in Europe around 2500 BCE. In his comprehensive review of this topic, Stuart Piggott reviewed a huge amount of evidence for the spread of the wheel and wheeled transport, from its invention to its ubiquitous use across Eurasia in the Iron Age.[31] The evidence comes from scratched images on pottery sherds and rocks and from the fragments of wooden wheels and carts, and also from the excavated burials of draft animals, sometimes with their harness. But even without the harness remains, if there are skeletons of animals buried in pairs it may be assumed that, in life, the animals had been used to draw a cart or chariot.

Burial of a pair of cattle from a third millennium BCE excavation in Hungary provides early evidence for the use of oxen in draft, while a pottery cup in the shape of a wagon from the same site shows that disk wheels were in use. Other third millennium BCE finds of wheels and axles have been excavated from waterlogged sites as far west as Switzerland and Denmark, and Piggott believed that diffusion of wheeled vehicles drawn by oxen occurred relatively fast across Europe from Mesopotamia in a few hundred years.[32]

During the second millennium BCE, horses were beginning to replace oxen for transport. They could be ridden, and they could also be harnessed to light wooden vehicles with spoked wheels, termed "chariots" throughout the ancient world, although their function was not always for hunting and war as it was with the Mesopotamian chariots. The new element in this new

Figure 17. Pottery cup in the form of a wagon, early third millennium bce from Szigetszentmárton. (Source: Stuart Piggott, *The Earliest Wheeled Transport from the Atlantic Coast to the Caspian Sea* [London: Thames & Hudson, 1983], 46. Reproduced with permission from the Hungarian National Museum.)

means of transport was speed, for whereas the heavy ox wagons traveled at something like 3.7 kilometers (2 miles) an hour, a horse chariot could travel at 38 kilometers (20 miles) an hour.[33]

Mary Littauer and J. H. Crouwel have argued that the horse-drawn chariot with spoked wheels was first invented in the Near East early in the second millennium BCE rather than being an import from Indo-European steppe tribes,[34] and, as Piggott believed, it is reasonable to assume that this great advance in technology then diffused westward into Europe. The adoption of the chariot was, however, not universal, and in some regions it took a long time to arrive; in Britain, there is no evidence of horse harness or chariot burials before the Iron Age (c. 500 BCE).[35]

THE BEGINNINGS OF NOMADIC PASTORALISM

The first farmers all over Europe probably lived as agro-pastoralists in small social and family groups cultivating a piece of land with cereal crops while owning a few scavenging dogs and pigs, and a few livestock, which would be sheep and goats and perhaps a cow. They would also supplement their food supply with milk and with hunted meat and fish or invertebrates from the sea. As described above, the milk would most likely have been fresh in central and northern Europe, but in the Mediterranean region, where sheep and goats were more common than cattle, it would most likely have been converted into by-products such as cheese (which does not require the presence of lactase for digestion), signified by the dominance of lactose malabsorption in southern Europeans today.

As the human population increased over the millennia, families would have become divided into the rich and the poor, with chieftains who owned big areas of land and large herds of livestock, while the rest of the population lived as peasants eking out a meager existence on small patches of land. In different parts of Europe and Asia and at different periods from the Neolithic to the present, both classes of people have had to resort to moving their herds over large distances in order to find sufficient grazing lands. In southern Europe and in mountainous regions, the livestock are moved seasonally between highland and lowland areas, which is termed "transhumance," while in other areas and in the steppes of northern Eurasia herds are moved periodically to different pastures around the region, which is termed "nomadic pastoralism." If, as in many areas, there is also cultivation of crops, the term used is "semi-nomadic pastoralism," as defined by Khazanov in his classic work, *Nomads and the Outside World.*[36]

Khazanov described the origins of nomadic pastoralism as going back to the "Neolithic revolution"—that is, to the origins of cultivation and animal husbandry—and where this food-producing economy led to adaptation to a different habitat, especially in arid lands, there would be a predominance of pastoralism over other forms of agriculture. This process took several thousand years, and it was only in the third and second millennia BCE (the Bronze Age) that pastoralism, in several different forms, spread throughout the Eurasian steppes. Khazanov links the final emergence of nomadism to the drying of the climate, which has been recognized for the second millennium BCE by paleoclimatologists.

Sevyan Vainshtein, in his study of the nomads of Tuva and their history, has written that southern Siberia and Mongolia are among the areas with the most ancient pastoralist traditions of the Old World, and that "the Przhevalsky [Przewalski] horse,[37] the Bactrian camel, the yak, and the northern reindeer survive here to this day; these are the wild species most closely related to the domestic animals herded in this region."[38] These herds, with the additions of sheep and goats, go back thousands of years in their pastoral migrations and survived through the many conquests of the land by, for example, the Scythians in the mid first millennium BCE, the Uyghurs in the eighth century CE, and the Mongol empire in the thirteenth and fourteenth centuries CE.[39]

It is difficult to imagine nomads of the Eurasian steppes not using the horse as a riding animal. Khazanov quotes P. S. Pallas, writing in the middle of the eighteenth century, as stating that a nomadic Kalmuck family (the western Mongols of Central Asia) required eight mares, one stallion, ten cows, and one bull to survive.[40] In other areas of the steppe, nomads have herded sheep, cattle, and camels, which they follow on foot and in carts drawn by oxen or horses. Throughout the Eurasian steppe, horses have been a pivotal animal for all nomads, providing transport, draft, milk, hair, and, at their death, meat, hides, sinews, and bones.

Domesticates in Ancient Egypt and Their Origins

CLIMATE CHANGE IN THE SAHARA FROM 18,000 TO 4,000 YEARS AGO

Toward the end of the last Ice Age and into the Holocene, the climate of North Africa went through dramatic alternations of wet and dry periods. From around 18,000 to 12,000 years ago there was a period of extreme aridity in the Sahara, which drove out all human inhabitants and most of the animals, except for where they could retreat to refuge areas in the eastern Nile Valley and on the northern coastline. Then, with the melting of the ice in the north, from 12,000 to around 8,000 years ago, the rains returned, and much of the Sahara became grasslands with abundant lakes. Hunter-gatherers and their prey moved back, and right in the middle of the Sahara at the site of Adrar Bous bone harpoons have been found that were used to kill hippopotamuses in the nearby lake.[1]

The transition of hunter-gatherers in North Africa to food production and pastoralism can be seen as a response to these extreme climatic changes in the huge area of the Sahara. For around 500 years, from 8,000 years ago, there was a dry period north of the Equator that was paralleled in the Levant at the end of the Pre-Pottery Neolithic B period. Following this, from 6,500 to 4,500 years ago, the climate became wetter again and led to the archaeological period known as the Neolithic Wet Phase throughout the Sahara and the Sahel, and was so called because the sites had pottery. By this time there was a nomadic pastoral economy based on cattle throughout the Sahara, but domestic sheep, which were present in the north around the Atlas Mountains, were rare in the central Sahara. The Neolithic Wet Phase predates the Predynastic period sites in Egypt (4500–3000 BCE[2]), where the remains of domesticated animals and cultivated cereals are common.

From around 4,000 years ago the climate began to get progressively drier again, until the Sahara became the hot, arid desert of today.[3] The pastoralists would have migrated to the desert margins and Nile Valley, where the Early Dynastic period of the Egyptian civilization was fully established.

SAHARAN ROCK ART

Throughout the Sahara, and especially in the north, there are abundant sites with the most wonderful rock art, the best-known images being from the Tassili N'Ajjer in Algeria. The earliest representations are of large wild animals such as elephant, giraffe, rhinoceros, and buf-falo, which are believed to have been figured by hunter-gatherers more than 7,000 years ago and possibly as early as 10,000 years ago. Later images show the introduction of domesticated cattle, sheep, and goats, then horses and chariots from around 3,000 years ago, and camels from around 2,500 to 2,000 years ago.[4]

There were no species of true wild sheep or goats that were indigenous in North Africa following the end of the Pleistocene, only the so-called Barbary sheep (*Ammotragus lervia*),

Figure 18. Hunters with dogs chasing a Barbary sheep (*Ammotragus lervia*). From a rock painting in the Wadi Teshuinat in the Libyan Sahara. (Credit: Joe Carnegie, copyright Libyan Soup.)

Figure 19. Cattle with their herders from the rock shelter of Jabbaren in Tassili. Both bulls and cows are shown with a variety of horn shapes, including some hornless, and mixed coat colors. (Source: Henri Lhote, The Search for the Tassili Frescoes, trans. by Alan Houghton Brodrick [London: Hutchinson 1959], 112, Plate III.)

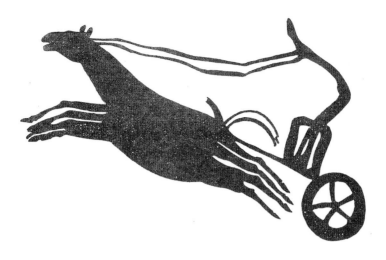

Figure 20. A chariot with driver and a pair of horses on a rock face from Wadi Teshuinat. (Source: Alfred Muzzolini, *L'Art Rupestre Préhistorique des Massifs Centraux Sahariens* [Oxford: Cambridge Monographs in African Archaeology 16, BAR International Series 318, 1986], 254.)

which, as shown in figure 18, were hunted with dogs. The earliest levels (tenth millennium before present) of the rock shelter of Uan Afuda in the Wadi Teshuinat have produced almost exclusive remains of the Barbary sheep. Later levels in the interior of the cave, dating to 8,000 years ago, have produced huge numbers of compacted coprolites (fossilized dung), indicating that the animals were penned and probably kept in a state of semidomestication. However, this phase of experimental domestication of the Barbary sheep did not last, and 1,000 years later they had been replaced by domestic cattle, sheep, and goats herded by nomadic pastoralists.[5]

ORIGINS OF DOMESTICATES IN NORTH AFRICA

Traditionally, it has long been claimed that the only animals to have been certainly domesticated in Africa are the cat, the donkey, and the Guinea fowl. Recent molecular studies, however, have partially altered this picture. It looks as though the cat was actually first domesticated in the Near East,[6] but the donkey (*Equus asinus*) did originate in Africa and is descended from two distinct subspecies of wild ass, the Nubian (*Equus africanus africanus*) and the Somali (*Equus africanus somaliensis*), both of which still inhabit northeast Africa, but they are critically endangered. The Nubian wild ass is distinguished from the Somali by the former's strong dark stripe down the shoulders and absence of stripes on the legs, while the Somali ass always has marked leg stripes and usually no shoulder stripe.[7] Both the leg and shoulder stripes are often present in the domestic donkey, which supports the genetic evidence that both wild subspecies are its ancestors.[8] There is pictorial evidence from prehistoric rock art and Roman mosaics that a third, separate species of wild ass inhabited Algeria, but it has long been extinct, and there is no osteological evidence of its existence.[9]

The Guinea fowl is an endemic African bird, with several wild species in West and East

Africa. The most common domesticated species is the helmeted Guinea fowl (*Numidea melea-gris*). It was known to the ancient Egyptians and spread to Greece in classical times.

Dogs, sheep, goats, and horses are all without any fossil ancestors on the African continent, so there can be no argument about their importation: dogs and horses from Eurasia, and sheep and goats from western Asia. The origin of domestic camels, pigs, and domestic cattle in North Africa is more enigmatic. The dromedary, or one-humped camel (*Camelus dromedarius*), is perfectly adapted to life in hot, dry sand-deserts, but there are no wild dromedaries in existence, and it is not known where they were first domesticated. Nor is it known if herds of wild dromedaries inhabited the Saharan region during the wet phases. The earliest Holocene remains of camels have been excavated from the Red Sea coast of Arabia, dated to 9,000 years ago.

Wild boar (*Sus scrofa*) inhabited the Mediterranean borders of North Africa and the Nile Valley until recent times, and it is not impossible that they were locally domesticated, as it is now assumed they were in many European and Asian localities, but there is no material or molecular evidence for this. On the other hand, it is equally possible that wild boar were imported by humans into North Africa at an unknown early period from Eurasia.

The origins of domestic unhumped cattle (*Bos taurus*) and humped cattle (*Bos indicus*), together with their pastoralist herders, have been well studied throughout Africa for many years, but there is still more to learn. The initial problem was whether unhumped and humped cattle were descended from a single species of wild cattle, the aurochs (*Bos primigenius*), or whether humped cattle were descended from a separate species that was endemic to India. Molecular analysis has now shown that the genetic distance between the two types of domestic cattle is so great that their ancestors must have become separated as breeding populations long before domestication took place. It is therefore certain that unhumped cattle are descended from *Bos primigenius* and that humped cattle are descended from a geographical race of aurochs that is found as fossils in India and has been named *Bos namadicus*.[10] This accords with differences in the skull morphology of the two domesticates, taurine and indicine, as reported by Caroline Grigson in 1980.[11]

Molecular analysis has also solved the next problem, which was whether African cattle were first imported from western Asia through Egypt or whether they were domesticated directly from the indigenous race of aurochs (*Bos primigenius*) that inhabited North Africa but became extinct in the prehistoric period. Geneticists have found that the mtDNA of living unhumped African cattle is different enough from Eurasian breeds to indicate that they have a uniquely African origin.[12] This leads to the conclusion that in the early prehistoric period, there were three centers of cattle domestication: western Asia, North Africa, and India.

The earliest African domestic cattle have been identified from archaeological sites in the western desert at the border of southern Egypt and northern Sudan. These sites in the Nabta Playa basin have been excavated since the 1970s and have revealed the increasing complexity of an elaborate culture of cattle pastoralists spanning from 11,000 to 6,000 years ago. This covers the wet phase of the early Neolithic period and the short arid phase around 8,000 years ago, with another wet phase following, until the final drying of the desert. In the earliest period there are few remains of cattle, and there has been controversy about their identification and domestic status, but by 8,000 years ago there can be no doubt that pastoralism was well established. The remains of imported sheep and goats become common, and the skeleton of a cow has been retrieved from a clay-lined and roofed chamber below a mound, dated to 7,500 years ago, while other cattle remains had been intentionally buried under tumuli. The excavators of the site, Fred Wendorf and Romuald Schild, have postulated that this site was a cultural center where the herders would have met for religious, political, and social functions.[13]

Figure 21. The skeleton of a cow buried under a tumulus at Nabta Playa, dated to around 7,500 years before present. (Source: Fred Wendorf and Romuald Schild, "Nabta Playa and Its Role in Northeast African Prehistory," *Journal of Anthropological Archaeology* 17 [1998]: 1110, Figure 5. Reproduced with permission.)

As Wendorf and Schild point out, it is tempting to see an affinity in this culture with the religion of ancient Egypt. During the Old Kingdom (2686–2181 BCE), cattle were central to the Egyptian civilization and were regarded as representatives of the gods. Hathor was a very ancient and most important cow goddess, going back to the Predynastic period (c. 4500–3000 BCE). She was depicted as a cow with a sun disk between her horns or as a queen with a sun disk and horns on her head.

It was not only in Nabta Playa, in the Western Desert, but in many sites in the central Sahara and in the north, archaeological excavation has revealed the presence of extensive cattle pastoralism in the early Neolithic and stretching into much later periods. One such population, which has been called a "civilization," comprised the people who lived in the Libyan desert south of the Roman occupation on the Mediterranean coastline and who came to be known as the Garamantes. By 1000 BCE these people were probably successful cattle pastoralists, but their heyday was from 500 BCE to 500 CE, when they developed an elaborate irrigation system for their crops and traded slaves and salt with the Romans.

The Garamantes are well known from their description by Herodotus, a Greek historian and writer who lived from 484 to 424 BCE. Herodotus was a great traveler, and knowledge about many ancient civilizations and peoples, especially those of ancient Egypt, comes from his detailed accounts. He wrote the following description of the Garamantes:

> Ten days journey from Augila [an oasis in southern Libya] there is again a salt hill and a spring; palms of the fruitful kind grow here abundantly, as they do also at the other salt hills. This region is inhabited by a nation called the Garamantians, a very powerful people. . . . In the Garamantian

country are found the oxen which, as they graze, walk backwards. This they do because their horns curve outwards in front of their heads, so that it is not possible for them when grazing to move forwards, since in that case their horns would become fixed in the ground. . . . The Garamantians have four-horse chariots, in which they chase the Troglodyte Ethiopians.[14]

This is a most interesting account for several reasons, but perhaps mainly because it shows that in c. 450 BCE these desert people had developed a distinct breed of cattle with an equivalent horn shape to that of the old English breed of longhorn, as in the early-nineteenth-century engraving in figure 22. The horns of longhorn cattle can grow forward, as described for the Garamantian cattle, and can prevent the animals from grazing. If this occurs today, the ends of the horns are typically cut off.

In a later passage Herodotus gives support to the modern view that pastoralism was so successful in the Sahara because the herders were able to use milk, and probably blood, as an essential component of their diet. He wrote:

> Thus from Egypt as far as Lake Tritônis Libya is inhabited by wandering tribes, whose drink is milk and their food the flesh of [wild] animals. Cow's flesh however none of these tribes ever taste, but abstain from it for the same reason as the Egyptians, neither do any of them breed swine. Even at Cyrêné, the women think it wrong to eat the flesh of the cow, honouring in this Isis, the Egyptian goddess, whom they worship both with fasts and festivals.[15]

Figure 22. An English longhorn cow from an engraving dated 1 July 1807 (unknown artist).

DOMESTIC ANIMALS IN ANCIENT EGYPT

The people of ancient Egypt were more involved, as a nation, with wild and domestic animals in cultural and religious ways than any other civilization before or since. Also, more is known about the interactions between humans and animals in Egypt than for any other ancient civilization, the reasons being the vast numbers of well-preserved paintings in the tombs and temples, as well as the contemporary written descriptions and the untold numbers of mummified animals that had been buried in tombs and sarcophagi.

There are many accounts of ancient Egypt and its animals by classical Greek and Roman authors, one of the earliest being Herodotus, who wrote a long and detailed book on every aspect of Egyptian life in c. 450 BCE, which has been divided by the translators into 186 short chapters. His book ranges over a great variety of topics—from the process of embalming human and animal corpses, to the structure of the pyramids, to diet, to the use of mosquito nets, and to the sacred nature of animals. Herodotus begins his discourse on animals with the comment that although Egypt was bordered by Libya, it was not a country abounding in animals, and for this reason all the animals that are there, domesticated or otherwise, are regarded as sacred and consecrated to the several gods. However, some animals were clearly considered to be more sacred than others, and those that were useful in killing snakes and rodent pests, such as cats, mongooses, sacred ibises, and hawks, were mummified when they died, and if one was killed intentionally the perpetrator was executed.[16]

The Egyptians do not appear to have made a clear distinction between wild and domestic species. It was more that some species were sacred and were never killed because they were useful; others, such as crocodiles, were sacred but could be killed; and others, including fish, sheep, and birds such as quails and geese, were killed for food.

To the accounts of Herodotus can be added those of another Greek historian, Diodorus Siculus, who lived 300 years later from 80 to 20 BCE. Although he may have copied some facts from Herodotus and other historians, Diodorus Siculus did visit Egypt, and at least some of his accounts were written from firsthand experiences. Like Herodotus, Diodorus Siculus wrote that there were animals that were useful and animals that were harmful, as well as some, like the crocodile, that were both harmful and useful. A few examples of his descriptions are given below:

> The Egyptians venerate certain animals exceedingly, not only during their lifetime but even after their death, such as cats, ichneumons [mongooses] and dogs, and, again, hawks and the birds which they call "ibis," as well as wolves and crocodiles. . . . The attendants of the animals cut up flesh for the hawks and calling them with a loud cry toss it up to them, as they swoop by, until they catch it, while for the cats and ichneumons they break up bread into milk and calling them with a clucking sound set it before them, or else they cut up fish caught in the Nile and feed the flesh to them raw.[17]

> The multitude of [crocodiles] in the river and the adjacent marshes is beyond telling, since they are prolific and are seldom slain by the inhabitants; for it is the custom of most of the natives of Egypt to worship the crocodile as a god, while for foreigners there is no profit whatsoever in the hunting of them since their flesh is not edible. But against this multitude's increasing and menacing the inhabitants nature has devised a great help; for the animal called the ichneumon . . . goes about breaking the eggs of the crocodiles . . . without eating them or profiting in any way it continually performs a service which, in a sense, has been prescribed by nature and forced upon the animal for the benefit of men.[18]

Most men are entirely at a loss to explain how, when these beasts [crocodiles] eat the flesh of men, it ever became the law to honour like the gods creatures of the most revolting habits. Their reply is, that the security of the country is ensured, not only by the river, but to a much greater degree by the crocodiles in it; that for this reason the robbers that infest both Arabia and Libya do not dare to swim across the Nile, because they fear the beasts whose number is very great.[19]

Cattle

The first domestic cattle to reach the Nile Valley may have moved there with their herders from the Sahara as the desert dried up in the Predynastic period. Cattle may also have been brought from the Levant, but however they reached Egypt, they soon became the most important livestock, and every aspect of their use on the land, and as suppliers of meat, dairy products, and material resources, is portrayed in huge numbers of realistic images on tomb paintings and on stone reliefs. Cows were milked, castrated oxen were used for plowing, and bulls were sacrificed to the gods.

At least four distinct breeds of cattle were developed during the 5,000 years of the Egyptian civilization. The earliest had long, lyre-shaped horns; then came short-horned breeds and a polled (hornless) breed. During the Eighteenth Dynasty (c. 1550–1307 BCE), there is the first pictorial evidence for humped zebu cattle, which must have been imported from Asia, either from the Levant or in boats across the Horn of Africa.

Figure 23. Inspection of the cattle on the estate of Nebamun. The upper row depicts the first attested portrayal of zebu with short horns and a marked thoracic hump, c. 1400 BCE; longhorned taurine cattle are depicted in the lower row. (Source: British Museum, No. 37976, "Wallpainting from a Theban tomb." Copyright Trustees of the British Museum, reproduced with permission.)

Sheep

There were two quite separate breeds of sheep in ancient Egypt. The earliest breed was a sheep that appears to have had a thin wool or hair coat rather than a heavy fleece, and with long, curving, corkscrew horns. During the Middle Kingdom (2040–1640 BCE), a new breed was introduced from western Asia, which had a heavy fleece and horns that curved round the side of the head. By the beginning of the New Kingdom (1550–1070 BCE), this breed had replaced the

Figure 24. The flock of sheep is being used to tread freshly grown seeds into muddy soil. From the tomb-chapel of the mastaba of Ti (no. 60) at Saqqara. (Source: Patrick F. Houlihan, *The Animal World of the Pharaohs* [New York: Thames and Hudson, 1996], 21, Figure 16. Wikimedia Commons, Credit: Einsamer Schütze.)

Figure 25. A stela dedicated to the god Amun. Two rams with heavy fleeces and recurved horns and wearing head-dresses are shown standing on shrines and shaded with large fans. Nineteenth Dynasty (1307–1196 BCE). (Source: Patrick F. Houlihan, *The Animal World of the Pharaohs* [New York: Thames and Hudson, 1996], 6, Figure 5. Reproduced with permission from Fondazione Museo Antichità Egizie di Torino.)

corkscrew horned sheep, and the ram of this breed, which had large, recurved horns, came to play an important part in the Egyptian religion as representative of the great Amun, chief god of Thebes.[20] A long row of ram-headed sphinxes flanks the entrance to the great temple of Karnak.

Huge numbers of rams were mummified, with the earliest known example of an embalmed sacred ram dating to the Early Dynastic Period (c. 3000–2650 BCE). Some were elaborately wrapped in bandages, decorated with headpieces, and placed in coffins in special burial places.

Goats

Like cattle and sheep, goats entered Egypt at an unknown early period of the Neolithic, and they probably flourished at the desert edges, where they could browse on woody shrubs. According to Herodotus, some Egyptians (the Mendesians) would not kill male or female goats because they were identified, as in Greece, with the god Pan, who was painted with the face and legs of a goat.

Pigs

According to Herodotus, pigs were not sacred and were considered to be unclean animals. If a man accidentally touched a pig, he immediately hurried to the river and plunged in with all his clothes on. Swineherds were forbidden to enter the temples, and they were forced to marry within their own community. However, pigs were sacrificed to Bacchus and the moon at certain full moons, and afterward their flesh was eaten. Poor people who could not afford live pigs made pigs of dough, which they baked and offered in sacrifice. Herodotus wrote that he knew why pigs were "detested at all seasons except at this festival," but he did not think it proper to mention it.[21]

Patrick Houlihan has given a concise description of the finds of pig remains from early sites, their absence from any mention in the animal portrayals of the Middle Kingdom (c. 2040–1640 BCE), and the apparent extension of the taboo in the time of Herodotus. Houlihan writes that there is no obvious reason for the taboo, but "the pig's legendary association with grubbing, dirt, and filth may have prompted its generally low status in the eyes of the ancient Egyptians."[22]

Camels

There are few pictorial records of the one-humped camel, or dromedary, in ancient Egypt and no word for the camel in the Egyptian language. This "ship of the desert" must have been known to the people, but perhaps it was not needed as a pack animal, for boats on the Nile could be used to transport heavy loads, and donkeys and cattle could be used for everyday transport and travel. Dromedaries were first brought into common use in trade routes across the Eastern Desert in the time of Ptolemy II Philadelphus, who ruled Egypt from 285 to 247 BCE.[23]

Donkeys

Early pictorial evidence for the donkey in Egypt comes from the Libyan palette. This is a flat stone carved on both sides with rows of cattle, donkeys, and gazelles on one side. This ceremonial palette is dated to the Predynastic Naqada III period (c. 3200–3000 BCE) and is believed to have been brought from Libya, which may provide evidence for the origins of the earliest donkeys. The wild ass (*Equus africanus*) evolved as an inhabitant of the arid regions of North Africa and the Sahara, and its distribution may not have extended into the fertile Nile Valley, so it is reasonable to assume that the first domestic donkeys were imported into Egypt. That they came from Libya is supported by Herodotus, who wrote on the wild animals in Libya of "asses, not of the horned sort [hartebeest], but of a kind which does not need to drink."[24]

The wild ass that is attested from several lines of evidence as having existed in Libya and Algeria became extinct at an early period, and so far it has not been possible to obtain a sample of its bone for mtDNA analysis. However, as Carles Vilà and colleagues have documented, their genetic data reveal that there are two separate groups of haplotypes to be found in modern breeds of donkeys, and they ascribe these two clades to the living subspecies of wild ass, *E. africanus somaliensis* and *E. africanus africanus*, both of which live in the eastern desert, nowhere near North Africa.[25] Patrick Houlihan has given a succinct description of the place of donkeys in the economy of ancient Egypt, where the thousands of years of overloading of these poor beasts of burden clearly began.[26]

Horses

For the early part of its 5,000-year history, ancient Egypt was protected from invaders by its geographical barriers. Cataracts on the Nile prevented invasion from south of Khartoum, while the Nile Delta and the Mediterranean protected the north. To the west, Egypt was enclosed by the Libyan Desert and to the east by the Red Sea. However, this isolation could not last, and toward the end of the Middle Kingdom the Palestinian people, known as the Hyksos, moved into the Delta and built their capital at Avaris, the modern Tell el-Dabaa. The Hyksos ruled Lower Egypt for a century but were finally expelled at the end of the Seventeenth Dynasty (c. 1640–1550 BCE).

Traditionally, it has long been believed that the Hyksos introduced the horse to ancient Egypt, and indeed there is no pictorial evidence for either horses or two-wheeled chariots before their arrival. However, sheep, goats, and dogs had certainly been imported from western Asia 1,000 years earlier, so it is inconceivable that the Egyptians did not have knowledge of the horse as a domestic animal at an earlier period than the Hyksos. And indeed there have been rare finds of horse remains on sites dated to earlier than the Hyksos invasion, the most notable being the nearly complete skeleton, excavated in 1958, from the fortress of Buhen on the second cataract of the Nile north of Khartoum, and dated to 1675 BCE by the excavator, Walter Emery.[27]

From the time of the Eighteenth Dynasty (1550–1307 BCE), horses and chariots are mentioned in inscriptions and are frequently portrayed in all forms of art. The horses always appear to be similar to the Arab breed of today, with light heads and fine limbs; they were given names and received the best possible care in the king's stables. Chariots have been recovered from a number of tombs, together with harness and bridle parts, and these have been described in detail by Mary Littauer and J. H. Crouwel, who found that the Egyptian chariots closely resemble those from the Near East, so it may be assumed that the Egyptians copied the design of their chariots from Asian originals.[28] Horses were not generally ridden in Egypt until Alexander the Great invaded the country in 331 BCE, followed by the Ptolemies, the Hellenistic family who ruled Egypt for 300 years from 304 BCE from their new capital of Alexandria. Perhaps the reason why horses did not become common in Egypt as early as elsewhere was because they were expensive and troublesome to keep and were not as useful as donkeys. They were therefore seen as high-status animals only to be owned by the pharaohs and the elite of society, and their only functions were for ceremonial occasions and for sport, much as their role has become in the Western world today.

Dogs

Nomadic hunters who moved into the Nile Valley more than 8,000 years ago probably had dogs with them, and dogs have been in the country ever since, as they have been in much of the rest of the world. As hunters turned to herding and their settlements grew into the ancient Egyptian civilization, many dogs became important individuals, given names and wearing elaborate collars. They were nurtured and loved as tenderly as in most human societies ever since, the only difference being that the Egyptian dogs continued to be cared for after death by being embalmed and mummified.

Although there are short-legged dogs in some tomb paintings that look quite different,

Figure 26. Mummified dog from a New Kingdom tomb in the Valley of the Kings (1550–1070 bce). Although he was probably owned by a royal person, he would not be distinguishable from the village dogs that are to be seen everywhere in Egypt today. (Source: Michael Rice, *Swifter Than the Arrow: The Golden Hunting Hounds of Ancient Egypt* [London: I. B. Taurus, 2006], 73, Figure 22. Credit: Bob Partridge, Editor, *Ancient Egypt*, reproduced with permission from Bob Partridge.)

the majority of the dogs pictured are slender, long-legged animals with fine heads, either with prick ears or lop ears, and often with tails curled over their backs. They look like the village dogs that are to be seen everywhere throughout Africa and the Middle East today. These dogs belong to the ancient greyhound type, which, like the Arabian horse, as a result of natural and artificial selection, evolved the physical characteristics necessary for survival in a hot, arid climate where they were often required to run at speed. In the tomb paintings of the pharaohs' hunting hounds, this greyhound type is given an extreme form of elegance, and the dogs are given the name *tjesm*.

During the final period of ancient Egypt, dogs were mummified in animal cemeteries in great numbers in connection with Anubis, Wepwawet, and other canine gods. Anubis is the best known of these gods from the regal, carved wooden model found in Tutankhamun's tomb. Patrick Houlihan has described this god: "Always depicted entirely black, a color of resurrection and rebirth, Anubis was the jackal-headed embalmer of the dead and protector of human burials in his role as lord and sentinel of the necropolis."[29]

Cats

The molecular evidence appears to prove that the first domestic cats were bred in the Near East, but there can be no doubt that they soon reached Egypt, and by the Middle Kingdom cats had become household pets and were revered for killing not only rodent pests but also snakes. Herodotus wrote detailed descriptions of the laws surrounding the treatment of sacred cats, and Diodorus Siculus recorded that while in Egypt in 59 BCE, he saw a mob of Egyptians demand, and apparently secure, the death of a man connected with a Roman embassy because he had accidentally killed a cat, and this despite the fear that the Egyptians felt for the Romans.[30]

Although the killing of a domestic cat was seen as a crime that could be punished by death,

Figure 27. Part of a votive limestone stela showing the two sons of the royal craftsman Nebre in adoration of a large cat. Nineteenth Dynasty (1307–1196 bce). (Source: Patrick F. Houlihan, *The Animal World of the Pharaohs* [New York: Thames and Hudson, 1996], 87, Figure 63. Reproduced with permission from Fondazione Museo Antichità Egizie di Torino.)

there was another side to the treatment of the cat as a sacred animal. Literally tons of mummified cats have been excavated from animal cemeteries, particularly at Bubastis, and X-rays of some of these mummies from the collection in the Natural History Museum, London, has revealed that the cats were nearly all quite young animals that had been killed by having their necks broken. It looks as if the cats were being specially bred in huge numbers only to be killed and mummified as votive offerings.[31]

Birds

As would be expected on the Nile River and in the surrounding wetland environment, a very large number of bird species, especially waterfowl, geese, and ducks, were present in ancient Egypt, probably in greater numbers than they are today. From the very many realistic and sympathetic tomb paintings it is clear that the birds were well known to the people and appreciated for their beauty as well as for their use as food, and little distinction was made between the wild and the domestic.

Many species of geese and ducks were painted being caught in the marshes, fed with corn, or being plucked and cooked. Quails are shown being netted in the fields or carried in a basket. Patrick Houlihan reproduces these and many other images of birds in ancient Egypt in a comprehensive and beautiful book.[32]

There are not many depictions of birds that can be confidently identified as domestic pigeons (descended from the rock pigeon) because of the difficulty of separating them from wild doves, but Houlihan illustrates a faience tile painted with a pigeon in flight from the Twentieth Dynasty (1196–1070 BCE).[33] It is not evident that pigeons were kept in cotes as they are in Egypt at the present day in huge numbers.

The common chicken, descended from the jungle fowl of India and southeastern Asia, was not imported into Egypt until the Ptolemaic period, although chickens are known from Mesopotamia much earlier. Diodorus Siculus refers to poultry in his remarkable account of how the Egyptians had learned to hatch their eggs by some process of artificial incubation: "The men who have charge of poultry and geese, in addition to producing them in the natural way known to all mankind, raise them by their own hands, by virtue of a skill peculiar to them, in numbers beyond telling; for they do not use the birds for hatching the eggs, but, in effecting this themselves artificially by their own wit and skill in an astounding manner they are not surpassed by the operations of nature."[34]

The Honey Bee, *Apis mellifera*

The honey bee is the only domesticated insect. Its tropical origins as an Old World species and its domestication have been described by Eva Crane, who stated that the first evidence for the use of hives is from ancient Egypt.[35] Recent genetic analysis of the mtDNA of ten subspecies from different geographic areas has been published by L. Garnery and colleagues, who suggest that the Middle East was the center for dispersal of the species.[36] According to Houlihan, beginning with King Den in the First Dynasty and until the last of the Ptolemies, Egyptian monarchs bore the title "He of the Sedge and Bee" as part of their royal title, and a hieroglyph of a bee was always used in the writing of this title.[37]

Domesticates of the Ancient Israelites, Assyrians, and Scythians

DOMESTIC ANIMALS IN THE OLD TESTAMENT

Politically, the most important development of the Levant in the early first millennium BCE was the establishment of the Israelite state.[1] Its history is told in the Old Testament, beginning with the Book of Genesis, perhaps around 1500 BCE, and ending with the Books of Ezra, Nehemiah, and Chronicles sometime before Alexander the Great in the fourth century BCE. However, the Books have no established chronology, and it is impossible to date any of them with any accuracy. In ancient Egypt, the period covered by the Old Testament may have begun with the Eleventh Dynasty (2040–1991 BCE) of the Middle Kingdom, and it ended with the Ptolemaic period. In archaeological terms, the time span of both the Old Testament and ancient Egypt extends from the Middle Bronze Age into the Iron Age.

The laws and many other descriptions in the books of the Old Testament give a remarkably detailed insight into the way of life of this Bronze Age civilization, from the Israelite kings with their horses and chariots in battle to the pastoralist herders with their flocks. Almost as much is written about domestic animals and the laws covering them as is portrayed in the images and hieroglyphs of ancient Egypt for the same period. In both regions, religion dominated the way of life of the people, but the traditions and rules of the two religions were extraordinarily different from each other. In ancient Egypt, useful animals, from cattle to cats to hawks, were protected by their sacred status. This prevented them from being killed, except by the priests under special circumstances, and there was the extraordinary custom of mummifying those animals that died or were sacrificed to the gods.

For the ancient Israelites, species of animals were not sacred, and the attitude of the people to the animal world differed from that of the ancient Egyptians, although most of the same domestic species and many of the same wild ones inhabited both regions. Even so, the Israelite priests kept as firm a hold on the people as did the Egyptian priests, as shown in the laws written in the Books of Leviticus and Deuteronomy. These laws distinguish between the clean and the unclean of almost every species of living animal that the Israelites were likely to come across, from birds of prey, chameleons, lizards, moles, tortoises (unclean) to beetles (clean and therefore edible). With so many species, large and small, terrestrial and aquatic, that were untouchable, did the countryside teem with wildlife?

Although there are lists of every species of animal mentioned in the Bible to be found on the Web, there has been little written about the interactions of people and their domestic animals, and for this the work of Canon Tristram has never been surpassed. In his book of

1889, Tristram claimed to have written "a description of every animal and plant mentioned in Holy Scripture," and he is still the original authority for references to the ethnozoology of the wild and domestic animals of the Bible.[2] Seventy years later F. S. Bodenheimer published his *Animal and Man in Bible Lands*, which ranges over literary sources, science, and folklore. It is an interesting book, which ends with chapters on the Israelite sacrifices and a review of Frazer's (1854–1941) analysis of biblical folklore in his renowned work, *The Golden Bough*.[3]

Dogs

Maybe because the ancient Israelites were not hunters, their attitude to dogs was always different to that of the ancient Egyptians, who valued their hunting and pet dogs very highly. Tristram states that there are forty mentions of dogs in the Bible, but they are all spoken with aversion. As at the present day, probably every Israelite town and settlement had its population of scavenging dogs who were useful for clearing debris, so they were tolerated but given no care or attention and lived as more or less "pariah" animals. However, the pastoralists did use dogs to protect their flocks from the indigenous wild wolves and jackals, although without affection, as shown in, "But now they that are younger than I have me in derision, whose fathers I would have disdained to have set with the dogs of my flock."[4]

Equids

Horses were of little use to the nomadic pastoralists in the relatively contained, hilly, and arid country of Palestine, and they are only mentioned in the earlier books of the Old Testament as owned by the kings, harnessed to chariots, and driven in battle, as described so magnificently in the Book of Job: "He paweth in the valley, and rejoiceth in his strength: he goeth on to meet the armed men. He mocketh at fear, and is not affrighted; neither turneth he back from the sword. The quiver rattleth against him, the glittering spear and the shield. . . . He saith among the trumpets, Ha, ha; and he smelleth the battle afar off, the thunder of the captains and the shouting."[5]

Domestic asses, descended from the African wild ass (*Equus africanus*), were the common riding animal in the Old Testament, but they were also highly valued for use by women, and white asses were an especial mark of rank and dignity: Abigail rode on her ass to meet David,[6] and Deborah addressed the judges, "Speak ye that ride on white asses, ye that sit in judgement."[7] According to Tristram, Baghdad was celebrated for its white asses in his day because they were more fleet than others.[8] Donkeys were also the common pack animals and were yoked to the plow, but the laws forbade an ox and an ass to be yoked together.[9]

Until recent times there was a subspecies of wild Asian ass, the Syrian wild ass, *Equus hemionus hemippus*, that was indigenous in the Near East, and archaeologists believed that this subspecies of ass was tamed and was represented on images and in the archaeozoology of early sites. However, osteological and genetic studies now reveal that none of the subspecies of Asian ass was ever domesticated, although the Syrian wild ass may have been interbred with domestic donkeys to produce crossbred equids that had hybrid vigor but were sterile.

The Syrian wild ass was the smallest of five subspecies of Asian ass, of which the other four still exist in the wild. They all differ from the African wild ass and from most domestic donkeys in having no shoulder or leg stripes. The last Syrian wild ass died out in a zoo in 1928, but in Tristram's time they were still living wild in small numbers, and he was not surprised that there are numerous references to wild asses in the Old Testament; for example: "Who hath sent out the wild ass free? Or who hath loosed the bands of the wild ass? Whose house I have made the wilderness, and the barren land his dwellings. He scorneth the multitude of the city, neither he regardeth the crying of the driver. The range of the mountains is his pasture, and he searcheth after every green thing."[10]

Camels

The domestic dromedary, or one-humped camel, must have been widespread by the time of Genesis, for Jacob gave to his brother Esau "two hundred she goats, and twenty he goats, two hundred ewes, and twenty rams. Thirty milch camels with their colts, forty kine, and ten bulls, twenty she asses, and ten foals."[11]

Camels were in common use for riding, as pack animals, and for plowing: "And when the queen of Sheba heard of the fame of Solomon, she came to prove Solomon with hard questions at Jerusalem, with a very great company, and camels that bare spices, and gold in abundance, and precious stones."[12] It seems that camels were even harnessed to chariots: "For thus hath the Lord said unto me, Go, set a watchman, let him declare what he seeth. And he saw a chariot with a couple of horsemen, a chariot of asses, and a chariot of camels."[13]

Cattle

According to Tristram, cattle were the most important animals in the agricultural economy of the ancient Israelites. They plowed the land, trod the corn, drew wheeled carts, and carried loads, and references to them in the Old Testament are innumerable.[14] Furthermore, they were especially important animals for sacrifice: "And Samuel said, What meaneth then this bleating of the sheep in mine ears, and the lowing of the oxen which I hear. And Saul said, they have brought them from the Amalekites: for the people spared the best of the sheep and of the oxen to sacrifice unto the Lord thy God."[15]

Unhumped taurine (*Bos taurus*) bulls, oxen, and cows with short horns were the common cattle of the Israelites, but bovid vertebrae with bifid neural spines have been identified from a few Bronze Age and Iron Age sites, and as these are commonly found on the skeletons of zebu, it may be that the Israelites also had a few humped cattle (*Bos indicus*) in their herds.[16] Wild water buffalo (*Bubalus arnee*) are represented in bas-relief on a few Mesopotamian cylinder seals, but Tristram claimed that there was no mention of either the zebu or the water buffalo in the Old Testament, nor were they present as domestic cattle in the Palestine of his day.

Goats and Sheep

Goats are adapted to browsing on scrub and brushwood while sheep graze on grass and short herbage, so the two species feed most efficiently on the same pasture and were herded together by the ancient Israelites as they still are in all arid countries today. From the earliest time, goats were milked, and the kids were the most popular meat for feasts, while lambs were kept into adulthood for the sake of their wool: "Go now to the flock, and fetch me from thence two good kids of the goats; and I will make them savoury meat for thy father, such as he loveth."[17] Goat-skin bottles were the usual way of carrying liquids; they were kept until worn out, were repaired, and are frequently mentioned, as in the Book of Joshua, "wine bottles, old, and rent, and bound up."[18]

After cattle, sheep were the most important animals in the Old Testament, and, according to Tristram, they are mentioned about 500 times, with their milk and wool, which could be white, being preeminent.[19] Rams' horns were used as flasks, and ram lambs were the usual animals of sacrifice. New breeds of sheep, some with an excessively fat tail and some with a mass of fat on the rump, must have been first bred by artificial selection during the Bronze Age somewhere in the Middle East, and these sheep then spread rapidly throughout the region.[20] The fat from the tail or rump was used for burning in lamps and for sacrifice: "And he shall offer of the sacrifice of the peace offering an offering made by fire unto the Lord: the fat thereof, and the whole rump, it shall he take off hard by the backbone . . . and the priest shall burn it upon the alter."[21] However, the laws forbade the eating of fat: "Ye shall eat no manner of fat, of ox, or of sheep, or of goat."[22]

Figure 28. A fat-tailed ram. "Syrian sheep." (Source: H. B. Tristram, *The Natural History of the Bible* [London: Society for Promoting Christian Knowledge, 1889], 138.)

Pigs

For the ancient Israelites, anything to do with pigs was forbidden by religious laws, as it is for Jews and Muslims today: the Old Testament reads that the pig, "though he divide the hoof, and be cloven-footed, yet he cheweth not the cud; he is unclean to you. Of their flesh shall ye not eat, and their carcase shall ye not touch,"[23] and the Qur'an states, "That which dieth of itself, and blood, and swine's flesh . . . is forbidden you."[24]

It has become commonly accepted that the reason why pork was ruled to be untouchable in the ancient world was due to the omnivorous diet of pigs, which can include the excreta of other animals and humans, and because they commonly have parasites that spread to humans. This may be so, but pigs in the biblical laws are among many other species of birds, mammals, and insects that seem to have been declared unclean because they are anatomically different from other clean species. Camels, for instance, are unclean because they "cheweth the cud but parteth not the hoof." Pigs are different from sheep and other livestock in having the opposite problem: they are cloven-hoofed but they cheweth not the cud.

DOMESTIC ANIMALS IN THE ASSYRIAN STATE

> The Assyrian came down like the wolf on the fold,
> And his cohorts were gleaming in purple and gold;
> And the sheen of their spears was like stars on the sea,
> When the blue wave rolls nightly on deep Galilee . . .
> And there lay the steed with his nostril all wide,
> But through it there rolled not the breath of his pride;
> And the foam of his gasping lay white on the turf,
> And cold as the spray of the rock-beating surf.

So wrote Lord Byron in his famous poem of 1815 on the siege of Jerusalem in 701 BCE by Sennacherib, king of the Assyrians, who ruled from 704 to 681 BCE. In this poem Byron accepted the Old Testament account (also supported by Herodotus) that the siege failed and the entire army died overnight: "Therefore thus saith the Lord concerning the king of Assyria, He shall not come into this city, nor shoot an arrow there, nor come before it with shield. . . . And it came to pass that night that the angel of the Lord went out and smote in the camp of the Assyrians an hundred fourscore and five thousand; and when they arose early in the morning, behold, they were all dead corpses."[25]

Contrary to this report that the entire army died overnight, in 1830 an Assyrian account in cuneiform script was recovered from excavations of Sennacherib's palace at Ninevah that claimed the siege on Jerusalem was a success. Sennacherib and other rulers before and after him did indeed attack and subdue many cities in all the surrounding states, and descriptions of them have been transcribed from the thousands of cuneiform tablets excavated from the palaces, of which Nimrud and Ninevah are the best known.

A great deal is known about ancient Assyrian history, but it is very difficult to disentangle and follow unless one is steeped in the archaeology of western Asia. On the other hand, for the archaeozoologist, a wealth of information is revealed on the stone bas-relief sculptures of the domestic and wild animals that were a major part of the economy of the Assyrian empire.

Figure 29. A city is captured in the reign of Tiglath-pileser III, c. 745 BCE, and the spoils are carried off. There are goats with very long horns, fat-tailed sheep, and ox carts. The four-wheeled "cart" on the left is a siege engine. (Source: Nicholas Postgate, "The Assyrian Empire," in *The Cambridge Encyclopedia of Archaeology*, ed. by Andrew Sherratt and Grahame Clark [Cambridge: Cambridge University Press, 1980], 187, Figure 26.1. Copyright Trustees of the British Museum, reproduced with permission.)

Figure 30. Mastiff hounds with their attendants in the royal park of King Assur-ban-apli, 668–627 BCE. (Source: Bas-relief sculpture in the British Museum [118915]. Note: Name of king as in Nicholas Postgate, *The First Empires* [Oxford: Elsevier-Phaidon, 1977], 140. Copyright Trustees of the British Museum, reproduced with permission.)

As in the land of the ancient Israelites, sheep were the most important herd animals and were pastured with goats; transport by donkeys was the basis of the trading networks, and oxen were also common draft animals. Horses were almost entirely held in the ownership of the army and the kings and were mostly used in chariots. Mules, the progeny of a female horse and a male donkey, were bred as baggage animals as they have been worldwide ever since.

Although mules are sterile, so they cannot breed directly, their hybrid vigor gives them more stamina and greater weight-bearing power than either of their parents.

As in ancient Egypt, the depiction of ceremony was all-important to the Assyrian kings, and the magnificent stone bas-reliefs on exhibition in the British Museum not only show this to its full extent, but the sculptures also show the chariots and types of the domestic animals in perfect detail. There are mastiffs that must have been bred as guard dogs as well as for hunting lions, which are shown being let out of cages only to be shot at by the king at close range from his chariot. The chariot horses are stallions, heavily decorated with plumes and bronze disks, and harnessed with bridle, bit, and yoke as described in detail by Mary Littauer and J. H. Crouwel.[26]

IN THE LANDS OF THE SCYTHIAN NOMADS

While the Israelites and Assyrians were building temples and palaces, and establishing their empires in western Asia with the aid of horse chariots, tribes of nomadic horsemen were expanding their ranges north of the Black Sea over the vast steppes of the Ukraine and southern Russia. Chief among these tribes, in the first millennium BCE, were the rich and powerful Scythians, and much is known about their way of life from the detailed eyewitness accounts that were written by Herodotus, who visited their lands in about 440 BCE. The following extracts are quoted from his very much longer descriptions to give a flavor of what he had seen:

> Having neither cities nor forts, and carrying their dwellings with them wherever they go; accustomed moreover, one and all of them, to shoot from horseback; and living not by husbandry but on their cattle, their wagons the only houses that they possess, how can they fail of being unconquerable, and unassailable even?[27]

> The manner of their sacrifices is everywhere and in every case the same; the victim stands with its two fore-feet bound together by a cord, and the person who is about to offer, taking his station behind the victim, gives the rope a pull, and thereby throws the animal down; as it falls he invokes the god to whom he is offering; after which he puts a noose round the animal's neck, and, inserting a small stick, twists it round, and so strangles him.[28]

> When a year is gone by [after the initial burial of a king], further ceremonies take place. Fifty of the best of the late king's attendants are taken, all native Scythians, . . . and strangled, with fifty of the most beautiful horses. When they are dead, their bowels are taken out, and the cavity cleaned, filled full of chaff, and straightway sewn up again. This done a number of posts are driven into the ground, in sets of two pairs each, and on every pair half the felly of a wheel is placed archwise;[29] then strong stakes are run lengthwise through the bodies of the horses from tail to neck, and they are mounted up upon the fellies, so that the felly in front supports the shoulders of the horse, while that behind sustains the belly and quarters, the legs dangling in mid-air; each horse is furnished with a bit and bridle, which latter is stretched out in front of the horse, and fastened to a peg. The fifty strangled youths are then mounted severally on the fifty horses.[30]

It might seem that Herodotus's description of the burial practices of the Scythians was written too long ago and is too fanciful to be true, but it has received remarkable support from the excavations of Scythian tombs. Burial mounds, or kurgans, have provided proof in excess of the ancient descriptions with the most exquisite gold ornaments, felt clothing, and horse gear, as well as the skeletons and mummified remains of sacrificed horses and humans. The most celebrated of the Scythian burials are at Pazyryk in Siberia; they were discovered and excavated in the 1920s by Sergei Ivanovich Rudenko, a Russian archaeologist who lived from 1885 to 1969. Because the tombs are in a landscape of permafrost, the frozen organic materials in them were preserved in nearly perfect condition. There were richly embroidered saddle blankets, woolen tapestries, leather coffin decorations of cocks and elks (moose), and a large number of the most elaborate horse trappings and ornaments.[31]

Two hundred years before Rudenko's discoveries, a wonderful collection of Scythian gold objects came into the ownership of Czar Peter the Great of Russia, following his edict that unusual antiquities should be sent to the court. This resulted in the collection in 1716–1718 of some 200 gold and silver objects that had been removed from ancient barrows in the center of the nomadic world of the Scythians in Siberia and beyond its frontiers.

The most important animal for the Scythian nomads and the one on which their lives depended was the horse, and this has remained true for the pastoralists of all the steppe lands until modern times. Not only were horses essential for transport over the vast regions of the steppes, but they also provided meat, leather, hair, mare's milk (from which butter and cheese

Figure 31. Horses and their Scythian riders resting under a tree. Gold belt plaque in the Siberian Collection of Peter the Great, acquired in 1716. (Source: Hermitage: Si 1727 1/61. From S. I. Rudenko, *Sibirskaya kollektsiya Petra I* [The Siberian Collection of Peter the Great]. SAI, vyp. D319. Moscow, Leningrad, 1962, Pl. VII, I.) Reproduced with permission from The State Hermitage Museum.)

were produced), and the all-important koumiss, which is a slightly alcoholic drink of fermented milk described by Herodotus and still made today. Apart from huge herds of horses that were owned by the Scythians, the ancient nomads pastured cattle, sheep, and goats, and they lived almost entirely on a diet of meat and milk products. Herodotus claimed that the Scythians grew corn, not to eat themselves, but for sale to the Greeks.[32]

The Scythian way of life, which Herodotus described in such detail, has, with some modification of the macabre burial practices, continued little changed in all the nomadic peoples of central and northern Asia until the present day. Herds of horses are still the basis of these tribes, with camels, yak, and reindeer in addition to cattle, sheep, and goats in the northern and most mountainous areas such as the Republic of Tuva, as described in the work of Sevyan Vainshtein.[33]

Seventeen hundred years after Herodotus wrote about the Scythians, the Venetian traveler Marco Polo (1254–1324) wrote about his twenty years of journeys through the Mongol Empire in the time of Kublai Khan (1215–1294), grandson of the great Genghis Khan (c. 1162–1227). Kublai Khan was the last of the great Mongol conquerors, and although he became emperor of China and founded cities, his rule was still based on nomadic traditions. Kublai built a huge palace in his city of Shan-Tu, where he had a park with game animals, a mew with more than 200 gyrfalcons, and a stud of 10,000 snow-white stallions and mares, whose milk could only be drunk by the royal family.[34]

It is this palace that is remembered in the famous poem by Samuel Taylor Coleridge (1772–1834), of which the first verse is:

In Xanadu did Kubla Khan
A stately pleasure-dome decree:
Where Alph, the sacred river, ran
Through caverns measureless to man
Down to a sunless sea.

In many respects, Marco Polo's descriptions appear like elaborate extensions of those of Herodotus, particularly the sections that relate to horses, the use of their blood and milk for food, and the elaborate postal service run by imperial messengers. Every twenty-five miles over the course of the postal service there was a "horse post" where the messengers could sleep, and 400 fresh horses were stabled there at the command of Kublai Khan.[35]

For the postal service in the frozen north, Marco Polo may have written the earliest account of travel by dog sledge:

There is a stretch of this country where no horses can go, because it is a land of many lakes and marshes and so covered with ice and mud and mire that no horse can go there. This bad tract extends for thirteen days' journey. At the end of each day's journey there is a posting station, where messengers crossing the country can find lodging. At each of these stations there are kept fully forty huge dogs, scarcely smaller than donkeys. These dogs transport the messengers from one station to the next, that is for a day's journey. . . . They have made sledges without wheels, so constructed as to glide over ice or mud or mire without getting too deeply embedded. . . . On these sledges they lay a bear skin and one of these messengers takes his seat on top. The sledge is then pulled by a team of six dogs.[36]

Domesticates in the Classical World of Greece and Rome

CRETE AND THE MINOAN CIVILIZATION

At the time of Genesis and the beginnings of the Israelite kingdom, the ancient Minoans on Crete, like the ancient Egyptians, were at the peak of their Bronze Age civilization. Like all the peoples of temperate Eurasia, 3,000 to 4,000 years ago, most of the Minoans were pastoralists who herded sheep, goats, and cattle; kept dogs; and harvested cereal crops. The elite of this civilization, however, were different. They lived in large, elaborate palaces (or temples) with walls that were painted with beautiful frescoes. The best known of these palaces is at Knossos; its ruins were excavated and restored by Sir Arthur Evans between 1900 and 1931.[1] For archaeozoologists, an all-important feature of the palace of Knossos are the painted images of the bull-leaping ritual, in which a young man is shown having somersaulted onto the back of a bull after grasping it by the horns.

It is not known whether the bull in this ritual was sacrificed at the end of the "game" or "dance," but it is often argued that the Cretan bull-leaping was the forerunner of the Spanish ritualized bullfights. Bulls were certainly integral to Minoan culture, and they are represented not only in the wall paintings but also on many artifacts, such as seals and rhytons (drinking cups), but perhaps most impressively, the Cretan bull has survived in the legend of the labyrinth and the minotaur, a mythical monster with a bull's head and a man's body. The minotaur was believed to live underground in the Cretan Labyrinth, which was an elaborate mazelike construction built for King Minos of Crete. The legend spread to mainland Greece, where the minotaur was killed by Theseus. The bull cult on Crete, its significance, and the possibility that the legend of the minotaur arose from the frequent earthquakes that hit the island have been discussed by Michael Rice.[2] There is no knowledge of whether the bulls lived as wild animals on Crete or whether they were stabled ready for the "games."

HOMER: THE *ILIAD* AND THE *ODYSSEY*

The myths, legends, and tales of gods and heroes that have survived from the time of the Minoans in Crete provide much information about the interactions of humans and their domestic animals throughout the history of the ancient Greeks, and the greatest and also the oldest of these tales are Homer's *Iliad* and *Odyssey*.[3] Homer is believed to have been an Ionian from the island of Chios whose great epic poems were written down very soon after the first invention of the Greek alphabet in the eighth century BCE. The *Iliad* is much more than a legend, and it is hard to believe that the story was not based on real-life incidents, albeit the theme is the Trojan War and the destruction of Troy, which actually occurred five centuries before Homer wrote his epic. For example, perhaps this was Homer's own horn bow:

> At once he took out his polished bow, made of horn from a leaping wild goat that he himself had once shot under the chest as it sprang down from a rock; he had lain in wait in a hide, and hit the goat in the chest, so it crashed on its back on the rock below. The horns growing from its head were sixteen palms long. These a bowyer skilled in hornwork had prepared and fitted into a bow, then smoothed the whole to a fine polish and capped it with a tip of gold.[4]

And Homer was surely used to the bleating of sheep waiting to be milked before being reunited with their lambs when he wrote: "But the Trojans, like ewes standing innumerable in a rich man's farmyard, ready to give their white milk, and bleating incessantly as they hear their lambs' voices—so hubbub rose from the Trojans throughout the breadth of the army."[5]

But most evocative of all Homer's writing about animals in the *Iliad* are his descriptions of chariots with their riders and horses, as in this passage about the chariot race held at the funeral of Patroklos:

> There is a dry stump of wood, oak or pine, standing about six feet out of the ground. The rain has not rotted it, and there are two white stones driven in on either side of it, at the place where the road narrows, and there is smooth going for horses round it—maybe it marks the grave of some man long dead, or was set up as a goal in earlier times. Well, swift-footed godlike Achilleus has made it the turning post now. You must cut in very close as you drive your chariot and pair around the post, and let your own body, where you stand on the well-strung platform, lean a little to the left of the horses. Goad your right hand horse and shout him on, and make sure your hands give him rein: and have your left horse cut close to the post, so that the nave of your well-built wheel seems to be just touching it—but be careful not to hit the stone, or you will damage your horses and smash the chariot.[6]

It is of interest that mules clearly had a very high value in Homer's time, for the second prize in this race was to be "a mare six years old and unbroken, pregnant with a mule foal."[7] It should also be noted that in neither the *Iliad* nor the *Odyssey* is there any mention of riding of horses, only of driving chariots.

The *Odyssey*, too, is filled with evocative and seemingly accurate descriptions of animals, as in the reply of Telemachus to the gift of horses offered to him by Menelaus:

> Horses I will not take to Ithaca. I'd rather leave them here to grace your own stables. For your kingdom is a broad plain, where clover grows in plenty and galingale is found, with wheat and rye and the broad-eared white barley; whereas in Ithaca there is no room for horses to roam about in, nor any meadows at all. It is a pasture-land for goats and more attractive than the sort of land where horses thrive. None of the islands that slope down to the sea are rich in meadows and the kind of place where you can drive a horse. Ithaca least of all.[8]

And there is a description of the wild goats on the island of the Cyclopes:

[T]here lies a luxuriant island, covered with woods, which is the home of innumerable goats. The goats are wild, for man has made no pathways that might frighten them off, nor do hunters visit the island with their hounds to rough it in the forests and to range the mountain tops . . . and this land where no man goes makes a happy pasture for the bleating goats.[9]

These goats, feral rather than wild, would have been the descendants of tamed goats that were taken to the Mediterranean islands from western Asia thousands of years earlier, as described in Chapter 3. Then there is the long description of the swineherd Eumaeus with his great numbers of pigs, also described in Chapter 3, but best known is the account of the old dog Argus, who recognized Odysseus on his return and "who no sooner set eyes on Odysseus after those nineteen years than he succumbed to the black hand of death."[10]

The *Iliad* and the *Odyssey* are often read as a history of the destruction of Troy in the Trojan War (which took place in the thirteenth century BCE), but these great works may equally well be read for their accounts of the daily life of people and their animals in all levels of Homer's society in the eighth century BCE.

AESOP'S FABLES

Two centuries after Homer, in the sixth century BCE, Aesop, a Greek slave, is believed to have written the large collection of fables that have been widely known ever since. They give another insight into the interactions between the people of ancient Greece and their animals—for example, this fable shows that domestic cats were already common in Aesop's time and also that they were expected to kill mice:

A cat was enamoured of a handsome youth and begged Aphrodite to change her into a woman. The goddess, pitying her sad state, transformed her into a beautiful girl, and when the young man saw her he fell in love with her and took her home to be his wife. While they were resting in their bedroom, Aphrodite, who was curious to know if the cat's instincts had changed along with her shape, let a mouse loose in front of her. She at once forgot where she was, leapt up from the bed, and ran after the mouse to eat it.[11]

ARISTOTLE ON ANIMALS

After another two centuries, Aristotle's writings on animals leave the realm of myth and legend and enter the real world as accurately as it could be understood and interpreted in his time. Aristotle lived from 384 to 322 BCE, and he is the most important source of information on the animal world of the ancient Greeks at the peak of the classical period.

Aristotle's investigations into zoology are compiled into a series of books known as *History of Animals*,[12] *Generation of Animals*,[13] and *Parts of Animals, Movement of Animals, Progression of Animals*.[14] He wrote about more than 500 wild and domestic species, with humans being treated in the same way as all other animals. The breadth of knowledge covered is so great that it is hard to believe that one man could have learned so many facts, especially when these books are considered in relation to the rest of his great works on philosophy, art, politics, and much else, all of which have remained part of Western culture to the present day.

According to Aristotle, the purpose of his *History of Animals* was to obtain information through investigation—that is, to ascertain facts about each kind of animal—and then, as a second stage, to find out the "Causes" of these observed and recorded differences. In *History of Animals* the parts themselves are described, for although this work is to some extent physiological, its main object was to deal with the anatomy of organisms. However, a great deal is also written about the general appearance of domestic animals and their husbandry—for example: "In Syria, the sheep have tails a cubit broad, and the goats have ears a span and a palm long and in some the ears meet below towards the ground; and the cattle, like camels, have humps on their shoulders. And in Cilicia the goats are shorn like the sheep in other places."[15] This is fascinating information for it reveals that around 350 BCE there were fat-tailed sheep in Syria and the breed of goats that still today has overlong lop ears, as well as humped zebu cattle. And in Cilicia (Turkey), there may even have been a breed of valuable cashmere goats, as today.

CENTAURS AND THE FIRST HORSE-RIDING

In the legends of classical Greece, centaurs were creatures who were half man and half horse. They were believed to have come from Thessaly (central Greece), and in one legend they were claimed to be the offspring of Centaurus, the son of Apollo; in another, they were the progeny of Ixion, a king of Thessaly who was seduced by a cloud in the form of Juno.

What is more likely is that a group of people from the plain of Thessaly were seen at a distance riding horses, and this unfamiliar sight of horses, which appeared to have the heads and

Figure 32. Fight between a Lapith and a centaur on the Parthenon frieze, 447–432 BCE. (Source: Photo of the South Metope XXVII of the Parthenon frieze in the British Museum. Copyright Trustees of the British Museum, reproduced with permission.)

trunks of men, were perceived as mythical beings. To us, the riding of horses, donkeys, and mules, let alone camels, cattle, and reindeer, is so familiar that it is hard to envisage a world in which humans never sat upon animals for transport. Yet it seems that riding animals did not replace driving them in carts and chariots until the second millennium BCE. After several hundred years, by the time of Aristotle and Alexander the Great, horse-riding had become ubiquitous, although the centaurs remained in Greek mythology. They were a favorite subject in the art of classical Greece, and the battle between the Lapiths and the centaurs is central to the marble sculptures in the Parthenon frieze in the British Museum.

ALEXANDER THE GREAT AND BUCEPHALUS

As a young man, Aristotle lived in Athens and spent twenty years studying under Plato. Plato died in c. 348 BCE, aged eighty-one, and then Aristotle went to live in Lesbos, where, at the request of King Philip of Macedon, he became tutor to Philip's son, the young Alexander the Great, who was owner of Bucephalus, one of the most famous horses in history. Knowledge about the adult life of Alexander is best known from the *Anabasis of Alexander*, a composite history written from contemporary sources by Arrian (Flavius Arrianus), a Greek historian who lived from around 95 to 175 CE.[16] Arrian wrote about the death of Bucephalus but not about how Alexander came to own him; this famous legend was recounted in *Plutarch's Lives*, first written in c. 120 CE.[17] In 344 BCE Philonicus the Thessalian offered the horse for sale to Philip II of Macedon for thirteen talents, but he appeared to be unsalable as he was vicious and unmanageable. Philip's son Alexander, however, claimed he could manage the horse and promised to pay for it himself should he fail to tame it. He spoke soothingly to the horse and turned it toward the sun so that it could no longer see its own shadow, which had disturbed it. Alexander successfully tamed the horse and galloped off with it.

Alexander the Great lived for only thirty-three years, from 356 to 323 BCE, and his short life was one long series of battles and conquests. By this Iron Age period horse-riding had become the paramount means of travel, and riders, armed with bows, made up cavalries for battle, although chariots were still in use. According to Arrian, in 334 BCE Alexander marched from Greece to the Hellespont (Dardanelles) with an infantry of 30,000 and a cavalry of over 5,000 horsemen. They crossed the strait in "160 triremes and a good number of cargo boats," with Alexander leading the way into Asia.

Having conquered his way through western Asia and Persia, Alexander reached the land that is now Pakistan, where the battle of Hydaspes was his last great campaign, and it was here that Bucephalus died, either from wounds or from exhaustion, in 326 BCE. The battle was fought on the banks of the Hydaspes River (the Jhelum, see figure 34) in the kingdom of Porus, whose son was also killed in the battle. Porus was claimed to have advanced on Alexander with 4,000 horses, 300 chariots, 200 elephants, and 30,000 infantry. No doubt these numbers are exaggerated, but even so, it is evident that the Indians had vast resources of manpower, horses, and elephants equipped for battle, and it is remarkable that Alexander won this battle. Afterward he founded two cities where he had crossed the river, Nicaea and Bucephala, in memory of his horse.[18]

XENOPHON ON HORSEMANSHIP AND HUNTING

A generation before the birth of Alexander saw the life and works of Xenophon, a Greek philosopher, soldier, and historian who lived from about 430 to 354 BCE. Xenophon was born into the aristocracy of ancient Attica (southern Greece, including Athens), and he studied under Socrates before joining the army of the Ten Thousand, who were Greek mercenaries hired by Cyrus the Younger to seize the throne of Persia. Cyrus was killed in the battle, and Xenophon led the starving army back to Greece.[19] However, it is not for his description of this dramatic journey that Xenophon is best known to archaeozoologists, but for his two detailed and authoritative works on horsemanship and on hunting.

Horsemanship and the Duties of a Cavalry Commander

The first part of Xenophon's treatise, which is on training an individual rider for horsemanship, is set out in twelve chapters that give such sound advice it is hard to realize they were not written in recent times but more than 2,000 years ago in the fourth century BCE.[20] The headings of the twelve chapters on horsemanship are:

Judgement of a colt
Of breaking and training colts
How to judge of a horse for riding
Attention necessary to be paid to a horse by its owner
Qualifications and duties of a groom
How a horse is to be treated
Mounting, riding and exercising a horse
How a horse is to be taught to jump and prepared for military service
How fierce and high-mettled horses are to be managed
Management of the bit and bridle
Teaching a horse his paces
A horseman's arms and armour

The last chapter ends with an instruction on throwing a javelin from horseback, and here it must be remembered that the ancient Greeks, like the Romans, had no knowledge of stirrups. Riders could only stay on their galloping horses by balance and by gripping with their thighs, so to throw a heavy weapon and hit its target was no mean feat: "If the rider advance his left side, at the same time drawing back his right and rising on his thighs, and launch his weapon with its point directed a little upwards, he will thus send it with the greatest force and to the greatest distance."[21]

In the second part of this treatise, the nine chapters on the duties of a cavalry commander give a fascinating insight into the training of cavalry horsemen in ancient Greece for battle and for public "shows" or religious festivities. In every admonition Xenophon's care for the welfare of the horses and their riders is of foremost importance, as on marching: "On marches, the commander of cavalry ought constantly to consider how he may give rest to the backs of his horses, and afford relief to the riders as they proceed, whether by riding at a moderate pace, or by dismounting and walking at a moderate pace."[22]

Xenophon's second treatise is on hunting, and in this he concentrates on the kind of dog to be trained for hunting, on the use of nets and snares, and on the hunting of hares, wild boar, and deer. Lions, leopards, lynxes, panthers, bears, and other wild beasts, he claimed, were hunted in foreign parts.[23] As with his treatise on horsemanship, the welfare of the dogs is given great attention, although the hunter's prey does not receive this consideration. For example, at the end of a hunt, "after having rubbed down the dogs, quit the hunting-field, stopping occasionally, if it be noontide in summer, that the dogs' feet may not become sore on the way."[24]

DOMESTICATES IN THE ROMAN WORLD

According to legend, the city of Rome was founded on 21 April 753 BCE, which is actually about the time of Homer, although this is an unusual comparison. The presumed building of the city is described in great detail in Plutarch's life of Romulus, written sometime before Plutarch died c. 120 CE,[25] while Varro (116–27 BCE), who was responsible for the calculation of the date, had commented more than 100 years earlier that "it was divine nature which gave us the country, and man's skill that built the cities."[26] As described by Plutarch, 22 April (as in the modern calendar) was taken as the day of the foundation of Rome because this was the date of the annual feast of Palilia, an ancient day of purification when a dog was sacrificed. As with all early societies, the lives of the Romans were strictly controlled by their religious rituals and sacrifices.

The city of Rome probably grew from a group of small villages, and for more than 400 years the Romans continued to live in agricultural communities, with Etruscans in northern and central Italy and Greeks on the southern coast and in Sicily. The Roman population was, however, steadily increasing, and they were taking control of the land, expanding their territory, and looking abroad for further conquests. The first of the three Punic Wars began in 264 BCE.[27] The third ended nineteen years later in 146 BCE and resulted in the total destruction of Carthage. There are no contemporary writings about the history of Rome before the third century BCE, but from that time onward there is a wealth of literature on all aspects of the Romans in peace and war. These Latin books, combined with the finds from innumerable archaeological excavations, provide accurate reconstructions of the Romans and their empire that are unparalleled in the ancient world. The basis of this empire was a successful agricultural economy run by huge numbers of slaves and the increasing use and development of iron for tools such as ox-drawn plows, which replaced the old wooden digging sticks.

Remains of all the domestic animals that are common at the present day are found in great numbers in Roman archaeological sites, but no osteological remains have ever been excavated of one species that is no longer domesticated or even in existence as a living species, this being the North African elephant, which was used in battles both against the Romans and by them. Wild elephants must have been present in great numbers in North Africa, and the Carthaginians, who dominated the western north coast of North Africa in the third century BCE, learned how to keep them in captivity and train them for warfare. Elephants were used extensively in the first Punic War and then famously by the Carthaginian king Hannibal in the second war when he crossed the Rhône in Gaul with twenty-one elephants in 218 BCE, which is described in detail by Livy (59 BCE–17 CE).[28]

The Carthaginians were finally beaten by the Romans in 146 BCE, and their country was

Figure 33. A Carthaginian coin with, on one side, a portrait of Hannibal or his father, Hamilcar, and, on the other side, a war elephant, identifiable as the African species by the large ear, domed forehead, and high rump. The Carthaginian elephants are presumed to have been the small African forest species, *Loxodonta cyclotis*, rather than the savanna species, *Loxodonta africana*. (Copyright Trustees of the British Museum, reproduced with permission.)

a prize very well worth having, for Polybius (c. 200–118 BCE) wrote about the fauna of North Africa in his *The Rise of the Roman Empire*: "the total of horses, oxen, sheep and goats which inhabit the country is so immense that I doubt whether an equal number can be found in all the rest of the world." And "Again who has not read of the great numbers and the strength of the elephants, lions and panthers of Africa."[29]

The Romans who wrote books for farmers and stockbreeders provided advice that is wonderfully detailed and sound. These authors could equally well have been writing for today's small farmers, apart from the instructions for their slaves and practices concerned with their religious beliefs, which today would be considered as superstitions. The earliest of the agricultural writers was Cato the Elder, who lived from 234 to 149 BCE. Cato was brought up on his father's farm, but at the age of seventeen he began a military career that lasted for twenty-six years and during which he fought in the second Punic War. He was noted for his cruelty to defeated enemies,[30] but his writings on agriculture show sympathy for the livestock and their welfare, although he is always most interested in increasing profit and is strict about the housing and rations for slaves. For example, "Clothing allowance for the hands: a tunic three and a half feet long and a blanket every other year. When you issue the tunic or the blanket, first take up the old one and have patchwork made of it. A stout pair of wooden shoes should be issued every other year."[31] Cato's instructions for the construction of farm buildings and farm management are more general in their content than those of the later writers Varro and Columella, but he is often quoted as an authority by them.

Marcus Terentius Varro lived from 116 BCE until he was ninety years old in 27 BCE. Varro wrote his work on farming, *Res Rusticae*, in his eightieth year, and it was addressed to his wife, Fundania, who had just purchased a farm. Its three books are on "agriculture proper, domestic cattle [oxen, equids, sheep, goats, pigs], and the smaller stock of the farm, such as poultry, game birds, and bees."[32]

To Varro goes the earliest record of breeding rabbits in a warren for eating. In his time the rabbit, *Oryctolagus cuniculus*, was a wild species of lagomorph that was restricted in its distribution to Spain and possibly northwest Africa. The Romans were responsible for its spread around Europe, as described by Varro: "To the third species [of hare] belongs the one which is native to Spain—like our hare in some respects but with short legs—which is called cony. . . . [T]he conies are so named from the fact that they have a way of making in the fields tunnels (cunniculus) in which to hide. . . . [A]s you were in Spain for so many years that I imagine the conies followed you all the way from there."[33]

As part of the economy of the farm, Varro describes the keeping and fattening for profit a great range of animal species, such as lampreys, snails, dormice, and poultry of all sorts, including peacocks and pigeons. His description of these could be about those in my own dovecote, which are continually invaded by feral pigeons from the town: "in a dove-cote there are usually two species of these [doves]: one the wild, or as some call them the rock-pigeon. . . . [T]he other species of pigeon is gentler and being content with the food from the house usually feeds around the doorstep. This species is generally white, while the other, the wild, has no white, but is variously coloured."[34]

The best-known and indeed the most experienced and comprehensive writer on Roman agriculture was Lucius Junius Moderatus Columella, who was born in Spain c. 4 CE. After a career in the army, he lived on farms in Italy and died in 70 CE. Columella wrote twelve books on farming, which he named *De Re Rustica* and addressed to Publius Silvinus. All these books and another short one on trees have been preserved, with books VI–IX being on the rearing of livestock and dogs, veterinary medicine, poultry, fish ponds, and bees.[35] The overriding impression given by Columella's writing, as indeed that of Cato and Varro, is of care for the individual animals with correct housing and feeding, although of course the knowledge of medicine known to these authors was often erroneous, as in: "It will be found best to cut the tails of puppies forty days after birth. . . . As a result, the tail never grows to an ugly length and (so many shepherds declare) rabies, a disease which is fatal to this animal, is prevented."[36]

It is not easy to identify breeds of domestic animals from their bone remains on archaeological sites, but the descriptions of farm animals by Columella show that breeds of livestock were well established in the Roman period, and there were local breeds in the different regions of Italy. For example, there were small white oxen in Campania; huge white oxen, as well as red oxen, in Umbria; and thickset, powerful oxen in Etruria and Latium.[37]

And for sheep, Columella writes:

> Our farmers used to regard the Calabrian, Apulian and Milesian as breeds of outstanding excellence, and the Tarentine as the best of all; now Gaulish sheep are considered more valuable. . . . While white is the best colour, it is also the most useful. . . . By their very nature black and dark brown sheep also, which Pollentia in Italy and Corduba in Baetica produce, are esteemed for the price which they command; Asia likewise provides the red colour which they call "erythraean."[38]

The common domestic poultry on the Roman farm were chickens, descended from the Asian jungle fowl (*Gallus gallus* and *Gallus sonneratii*) and the African Guinea fowl (*Numida meleagris*). Columella advised that 200 fowls was the largest number that should be in the care of one person, and "an old woman or a boy should be set to watch over them." The hens should be the largest available and have red or darkish plumage and black wings, although black all over would be best. White hens should be avoided as they tended not to be prolific, and they were often carried off by hawks and eagles.[39]

The best-known and most prolific Roman writer on the animal kingdom was Pliny the Elder, a contemporary of Columella, who was born in 23 CE and died on 25 August 79 CE in the eruption of Mount Vesuvius, which destroyed the cities of Pompeii and Herculaneum. In the thirty-seven books of his *Natural History* Pliny aimed to cover the entire field of knowledge, with his writings on land animals in Book 8.[40] Much of the information is about wild species and lies in the realm of fantasy, but Pliny does distinguish between the wild, the tame, and the domestic; for example: "Hares are seldom tamed, and yet they cannot properly be called wild animals; indeed there are many species of them which are neither tame nor wild, but of a sort of intermediate nature; of the same kind there are among the winged animals, swallows and bees, and among the sea animals, the dolphin. Many persons have placed that inhabitant of our houses, the mouse, in this class also."[41]

In the time of Columella and Pliny, the Roman Empire was still at its zenith, but it was not to last. As the population increased, agricultural land grew progressively less fertile, and the forests over the whole Mediterranean region were cut down for shipbuilding and for fuel. This deforestation caused climate change with reduced rainfall, as it is doing in many parts of the world today, and the lands in North Africa that for centuries had provided grain and animals, especially horses, were becoming deserts. The Roman Empire slowly declined until the rule of Rome finally collapsed in 476 CE. Forty-seven years before this, on 26 March 429, in the eastern empire, Theodosius II attempted to reorganize and restore the laws of Rome. He announced to the Senate in Constantinople his intentions to codify all the laws, so twenty-two scholars worked from 429 to 438 CE to assemble what was to become the Theodosian Code. They produced sixteen books containing more than 2,500 constitutions that had been issued between 313 and 437 CE.[42]

Presumably because there was a shortage of draft horses and mules, many of the laws relate to control of their numbers and welfare, for without these essential pack animals the entire transport and military systems of the empire would have collapsed. For example: "We ordain that only one thousand pounds of weight may be placed upon a carriage,[43] two hundred pounds on a two-wheeled vehicle, and thirty pounds on a posthorse, for it appears that they cannot support heavier burdens. Eight mules shall be yoked to a carriage in the summer, of course, but ten in winter."[44] And because of the shortage of African horses: "We command that after the present twelfth year of the indiction the military horses paid as regular taxes of the diocese of Africa shall be commuted into money payments."[45]

Compassion and welfare for domestic animals is a theme that runs through all the writings of the Classical Greeks and Romans and is in marked contrast to the Roman treatment of wild animals in "sport" and human beings who had the misfortune to be defeated in battle or belonged to the lower orders of society. In an empire that was failing to feed its inhabitants and was enmeshed in bureaucracy and taxation, there was still room to provide for retired chariot horses: "We decree that provender from the fiscal storehouses shall be furnished to the Palmatian and Hermogenian horses when they have been weakened by their lot as contestants in the chariot races, either through the uncertainty of the race or by their number of years or by some other cause."[46]

Domesticates in Ancient India and Southeast Asia

ALTHOUGH A NOTABLE NUMBER OF EAST ASIAN DOMESTIC ANIMALS BELONG TO DIF-
ferent species from those of Europe and western Asia, the development of human societies, since the end of the Pleistocene period 10,000 years ago, has followed much the same course in the East as in the West. However, as in other parts of the world, particularly in Africa, the chronological sequence of cultural systems has not been straightforward in the East, and in quite recent times hunter-gatherer communities have often been found to subsist on their own or together with livestock herders. The site of Langhnaj in Gujarat (northern India) exemplifies this, for although the site may be dated as recently as about 2,000 years ago, the faunal remains consisted only of wild species including rhino, deer, nilgai, and wild boar. And the main artifacts were microlithic stone flakes, which in the West are the principal tools of Mesolithic hunter-gatherers that are dated from the ninth millennium BCE.[1]

The earliest evidence for the cultivation of plants and the herding of domestic livestock in South Asia has been excavated from the site of Mehrgarh in the region of Balochistan (Pakistan; see the map in figure 34). Over the many seasons of excavation of this well-known site, the animal remains have been identified and documented by Richard Meadow. More than seven periods have been revealed by excavation in the mounds and underlying strata, with the earliest level predating 6500 BCE. In this Pre-Pottery Neolithic level there were the remains of only wild animals, including wild sheep, goats, gazelle, wild ass, deer, wild cattle, water buffalo, and nilgai. As this period (1A) progressed through time, and based on the increase in numbers of cattle and sheep in the later levels, combined with a reduction in the size of the bones, Meadow identified the cattle as locally domesticated. He interpreted the sheep as possibly domesticated from a local form, which was later replaced by West Asian domestic sheep, while domesticated goats were present in all levels from the earliest times, but they were not necessarily of local origin. In addition, Meadow argued that the zebu (*Bos indicus*, descended from *Bos namadicus*, the Indian form of *Bos primigenius*) was present and had its earliest domestication at Mehrgarh.[2]

By the end of the third millennium BCE the herding of domestic zebu cattle, water buffalo, goats, and sheep characterized all the prehistoric sites in the region of the map in figure 34, but although all these species would have provided meat, it is not known if they were milked or if wool or hair was taken from the sheep and goats. However, Meadow believes that, from their large size, the sheep from Harappa (see the map in figure 34) could have been a specific breed, and they may have been wool-sheep, for which the presence of spindle whorls provides indirect evidence. Silk threads were also found at Harappa, together with beads or bangle fragments.

By the middle of the second millennium BCE, three species of transport animals had also been introduced to the sites from the northwest: Bactrian (two-humped) camels, donkeys, and horses.[3] By this period, the human population of the region had greatly expanded, probably as a result of the successful cultivation of crops and the herding of livestock. People had spread in great numbers into the Indus Valley, and at least five large urban centers had been established; in addition to the famous sites of Mohenjo-Daro and Harappa, there was Rakhigheri (Haryana), Ganweriwala (Cholistan area of southern Pakistani Punjab), Dholavira (Kutch, Gujarat), and perhaps Pathani-damb (foot of the Mula River, west Kachi Plain, Pakistan), and there may be more. Harappa gives its name, Harappan, to the archaeological industry found on all these surrounding sites. The Harappan period extends from the Early Bronze Age through the Late Bronze Age, with the Early Harappan dating from 3300 to 2600 BCE, the Harappan sensu stricto from 2600 to 1900 BCE, and the Late Harappan from 1900 to 1300 BCE.[4]

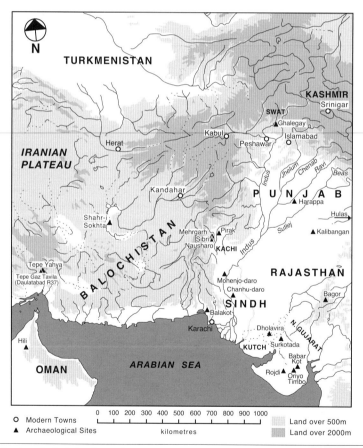

Figure 34. Map of archaeological sites in northwestern South Asia. The prehistoric sites are marked with a solid triangle. (Source: Richard H. Meadow, "The Origins and Spread of Agriculture and Pastoralism in Northwestern South Asia," in *The Origins and Spread of Agriculture and Pastoralism in Eurasia*, ed. by David R. Harris [London: UCL Press, 1996], 394. Reproduced with permission from Richard Meadow.)

Figure 35. Zebu bull on a steatite seal from Mohenjo-Daro. (Reproduced with permission from Images of Asia.)

Figure 36. Wild water buffalo bull on a steatite seal from Mohenjo-Daro. (Reproduced with permission from Images of Asia.)

Figure 37. *Rhinoceros unicornis* on a steatite seal from Mohenjo-Daro. (Reproduced with permission from Images of Asia.)

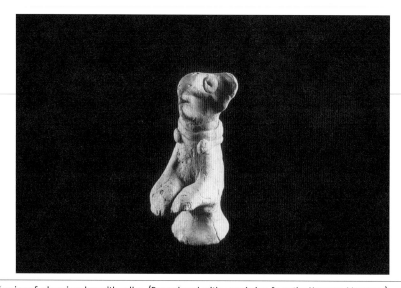

Figure 38. Figurine of a begging dog with collar. (Reproduced with permission from the Harappa Museum.)

Built as early as 4,500 years ago, many hundreds of years before any of the Roman cities, Mohenjo-Daro was one of the oldest cities in the world, and it rivaled anything that the Romans could build. Great numbers of houses of varying sizes were built along a grid of narrow streets. The houses were built of bricks with flat roofs of wood, and almost every house had a washroom with a chute that gave access to the main drains in the street outside. There must have been an elaborate system of trade in the city, and one relic of this remains in the numerous steatite (soapstone) seals that are believed to have been integral to this trade. Each

Harappan seal has an animal pressed into the stone as well as an inscription in an unknown script (see figures 35–38).[5] The seals illustrated here all date to the Harappan period (2600–1900 BCE).[6]

PASTORALISM AND CLIMATE CHANGE IN THE DECCAN PLATEAU

By the time that Mohenjo-Daro and Harappa were well established as urban centers 3,000 years ago, small farming communities were prospering in the favorable climate of the river valleys of the Deccan Plateau in southern India. The remains of one of these communities were studied in thirteen seasons of excavation in the 1960s and 1970s at the site of Inamgaon on the banks of the river Ghod in Maharashtra near Pune. Over 900 years, from 1600 to 900 BCE, more than fifty small mud houses were built at the site, but around 1000 BCE people began to leave this village and others in the region. In the early levels of the site domestic cattle bones greatly exceeded those of hunted animals, including blackbuck (*Antilope cervicapra*), which also outnumbered sheep and goats,[7] and the large numbers of animal bones indicate that meat made up a high proportion of the farmers' diet. By the ninth century BCE, the remains of hunted gazelle and antelope increased, while during the eighth century BCE all traces of habitation disappeared from the site. From these data Madhukar Keshav Dhavalikar deduced that the inhabitants of Inamgaon and other village sites in the northern Deccan were turning to nomadic sheep and goat pastoralism because the land could no longer sustain their crops. The reason was the drastic change in climate with much lower rainfall during the first half of the first millennium BCE, leading to greatly increased aridity and frequent droughts.[8]

DOMESTICATES THAT ORIGINATED IN ANCIENT INDIA AND SOUTH EASTERN ASIA

Domestic Yak, *Bos grunniens*

Wild yak, *Bos mutus*, with their stocky bodies and very long, thick coats, are adapted for life in the freezing cold of the high mountains of the Himalaya, and the domesticated form, *Bos grunniens*, differs little from the wild form. As a pack animal and provider of milk, the yak enabled humans to colonize the mountainous regions of Tibet and Nepal, and the domestication of this highly specialized bovid is of ancient origin. Yaks have always been important in the mythology of Tibet, and the tail, which is enormous with very long hair coming from its root, has been highly valued since antiquity and traded far from its origin. In China, the tail was traditionally dyed red and used as a decoration, while yak tail fans are a religious symbol in some Indian and Buddhist ceremonies.[9]

Zebu, or Humped Cattle, *Bos indicus*

The zebu, or humped cattle, of India and the East differ markedly in appearance from the Western unhumped taurine breeds, and genetic studies have shown that they do have a different and ancient origin. Humped cattle have a long, narrow skull; pendulous ears; a heavy dewlap; long legs; and a muscular or muscular-fatty hump on the shoulders. They are often light in color and are well adapted to hot climates. Although zebu, or indicine cattle, and taurine cattle share many characteristics in common and they will freely interbreed, it is now recognized from their anatomical distinctions,[10] but even more from their molecular distances, that they had quite different genetic origins. In 1994 Ronan Loftus and colleagues found that the divergence of the two mtDNA clades of taurine and indecine cattle was so great that they must have separated at least 200,000 years ago. This, of course, does not mean that these two lines of cattle were domesticated this long ago, which would be impossible, but that their ancestors, the European aurochs, *Bos primigenius*, and the Indian form, *Bos namadicus*, had separate evolutionary paths from that period.[11]

From ancient times, zebu cattle have been used throughout the subcontinent as draft oxen and for milk products, and the dung is very important for use as a fertilizer and as a fuel and building material. Today, beef is eaten in some communities, but the cow has always been a sacred animal in the Hindu religion and is never killed.

Figure 39. Zebu draft cattle. (Copyright Rebecca Jewell, reproduced with permission.)

Water Buffalo, *Bubalus bubalis*

The wild ancestor of the domestic water buffalo, *Bubalus arnee*, still survives in parts of India and Southeast Asia, but it is an endangered species. There are two types of water buffalo: the swamp buffalo of China and eastern Asia as far south as the Philippines, and the river buffalo of India and western Asia. It is possible that swamp buffalo, which are essentially draft animals, were first domesticated in the rice-growing areas of southern China or the Indo-Chinese peninsula. River buffalo, which are better producers of milk, may have been domesticated as early as the humped zebu cattle, but as their name implies, in wetland areas rather than in the dry plains and hills. In a study of the molecular biology of water buffalo, J. S. F. Barker and colleagues have shown that the genetic distances show a clear separation of the swamp and river types, which they estimated had diverged at least 10,000 to 15,000 years ago.[12]

Gayal, or Mithan, *Bos frontalis*

The gayal, or mithan, as it is also named, is a little-known bovid that may have been bred from the gaur, *Bos gaurus*, which is the largest of all living wild bovids. Gaur are forest cattle found as a rare animal in India and Southeast Asia. Gayal are maybe better termed as "exploited captives" rather than as domesticated cattle, and nothing is known of their ancient history. Gayal live as free-ranging forest browsers with the tribal peoples of eastern India, Bangladesh, and Myanmar (Burma), who control them with the provision of salt and only kill them as sacrificial animals.[13]

Figure 40. Domestic water buffalo in a tribal village in Sulawesi. (Copyright Galen Frysinger, reproduced with permission.)

Bali Cattle, *Bos javanicus*

Bali cattle are indigenous to the islands of Indonesia. They are descended from banteng, *Bos javanicus*, another rare and endangered wild species of cattle that live only in three wildlife parks in Java. Bali cattle are fully domesticated and are used for draft and for meat, but nothing is known of their ancient history.

Asian Elephant, *Elephas maximus*

Although Asian elephants can breed in captivity, it has been usual throughout history for all the thousands of elephants that have been trained for draft work, war, or ceremony to have been caught in the wild, either in stockades, with nooses, or in prepared pits. This is because it is uneconomic to breed elephants, as the gestation period is nearly two years, and the young elephants could not work until they were about fourteen years old. Therefore the Asian elephant has to be described as an exploited captive rather than as a true domesticate that has been subject to artificial selection.

Possibly the earliest evidence for the use of elephants is on seals from Mohenjo-Daro, dated to about 2500 BCE. Several of these may show the drawing of a rug on the elephant's back, including that in figure 41, which, if really a rug and not just the line of the shoulder muscle, would show that the elephant could not have been living as a wild animal.

Sixty years before the beginning of the Punic Wars, in which the Carthaginians used African elephants in battles against the Romans, Alexander the Great fought the famous battle of the Hydaspes River (Jhelum) against the Indian king Poros, whose main defense was his line of trained elephants. After traveling with his Macedonian army across the thousands of miles of the Middle East, in the late summer of 326 BCE Alexander arrived in the north of India and had crossed the Indus River.

Figure 41. Seal from Mohenjo-Daro (2500 BCE) showing an Indian elephant, possibly with a rug on its back as suggested by Zeuner (1963, 292). (Image courtesy of Richard H. Meadow.)

According to Arrian, who wrote using contemporary accounts, Poros advanced on Alexander with about 4,000 horses in his cavalry, 300 chariots, 200 elephants, and 30,000 infantry. The elephants were placed on the front line so as to terrify Alexander's cavalry. This they did, but the problem with elephants in battle is that they will move backward just as readily as forward, and they also very easily get out of control and charge about randomly. Although Alexander outmaneuvered Poros and won the battle, it was the end of his Indian campaign, and it was where his favorite horse, Bucephalus, died.[14] Elephants were never again used in major battles in India, but Zeuner claimed that one or more Indian elephants were actually taken to Carthage for the wars against the Romans, and it was from them that the Carthaginians learned to use the North African elephants. Some Indian elephants were also exported to defend Antiochus III at the battle of Raphia in 217 BCE against Ptolemy IV of Egypt.[15] The Indian elephant has remained a symbol of power to the present day and is still used in Southeast Asia for heavy draft work and for religious and state ceremonies.

Wild elephants are nomadic and require water and a very large area of land over which to graze and browse. Today, with the ever-increasing human population, there is less and less land for both African and Indian elephants to inhabit, and although individual trained elephants in Asia are often loved and given the best possible treatment, there is serious conflict with wild elephants who become crop-raiders when they leave the protection of the wildlife parks, which nowadays are nearly always the only places where they can live freely. Their escape from the protected areas often results in devastating economic losses to farmers, and death to both humans and elephants. In Asia, conflict between humans and elephants is particularly serious in the heavily populated Indonesian island of Sumatra, where Simon Hedges and Donny Gunaryadi have investigated various ways in which elephants can be deterred from raiding crops and invading agricultural land. However, the future is bleak for both Asian and African elephants in the wild.[16]

Figure 42. Asian elephants return home after a day's work, Elephant Nature Park near Chiang Mai, Thailand. (Copyright Agnes Arnold-Forster, reproduced with permission.)

Figure 43. Young Asian elephants sparring in the Elephant Nature Park near Chiang Mai, Thailand. (Copyright Agnes Arnold–Forster, reproduced with permission.)

Domestic Pigs, *Sus domesticus*

The relationship of pigs with humans has been wider and more complex than that of any other species of animal other than dogs, and the interweaving of the many different cultural attitudes to wild and domestic pigs can be traced through their history from Europe to Japan to New Guinea. And the wild boar, *Sus scrofa*, which has been the principal ancestor of the different breeds of domestic pigs over the whole of their vast range, is still extant. However, there are a great many different geographical races, or subspecies, of wild boar, with those in the East being morphologically distinct from *Sus scrofa scrofa* in the West. Until recently, the only way of assessing whether the remains of pigs from an archaeological site were likely to represent pigs domesticated from the wild boar of that region was with direct comparison of the ancient and modern bones and teeth. This never produced very satisfactory answers, but with the advancement of genetic studies, the new science of phylogeography has greatly altered the picture of where domestication of the pig has taken place. Although molecular studies still support the osteological evidence for two main centers of domestication in Europe and East Asia, the question now seems to be, "where has it not taken place?"[17]

All Eurasian pigs belong to the genus *Sus*, but the races of north India and the islands of Southeast Asia are much more complicated in their history than the single subspecies of wild boar, *Sus scrofa scrofa*, in the West. The molecular findings of Greger Larson and colleagues have shown that this diversity of races results from the initial evolution of the genus *Sus* in the Malaysian peninsula and the islands of the Indonesian Archipelago (see figure 44). The genus spread

Figure 44. Map of Southeast Asia and the Indonesian Archipelago. Reproduced with permission from The Free World Academy.

first into India, dividing into the Indian wild boar, *Sus scrofa cristatus*, and the pygmy hog, *Sus salvanius*, then into continental East Asia, and then into North Africa and western Europe.[18]

There are four defined species within the genus *Sus* that are living today on the Indonesian islands: bearded pig, *Sus barbatus*; Sulawesi warty pig, *Sus celebensis*; Javan warty pig, *Sus verrucosus*; and babirusa, *Babyrousa babyrussa*. These species of pigs have all been indigenously domesticated, interbred with *S. scrofa* races of domestic pigs, and moved around between the islands for thousands of years. Therefore it may be that Colin Groves, in his *Ancestors for the Pigs*, combined with his colleagues' molecular studies, has achieved as much as can be done to disentangle the individual histories of these races of pig.[19]

The pig and the dog were the earliest species to be domesticated in China, most probably both for their meat, and in 1969, according to Epstein,[20] there were 100 breeds of domestic pigs in China. These pigs are very different in their appearance and in the way their fat is stored from traditional European breeds, and a molecular study by Yong-fu Huang and colleagues has shown that the Eastern and Western domestic pigs are descended from separate maternal origins, Asian and European wild boars, respectively, which could have separated from a common ancestor 280,000 years ago.[21] These results have been refined by Greger Larson and colleagues, who suggest that domestic pigs were prevalent in both north and south China by 8,000 years ago and that modern Chinese domestic pigs are the direct descendants of the first domestic pigs in this region.[22] However, Chinese and Western breeds of pig will freely interbreed, and, from the end of the eighteenth century onward, Chinese pigs were

Figure 45. Species of pigs within the genus *Sus*: (a) different geographical races of wild boar, *Sus scrofa*; (b) pygmy hog from northeastern India, *Sus salvanius*; (c) bearded pig, *Sus barbatus*; (d) Javan warty pig, *Sus verrucosus*; (e) babirusa, *Babyrousa babyrussa*. (Source: Colin P. Groves, "Current Views on Taxonomy and Zoogeography of the Genus *Sus*," in *Pigs and Humans*: 10,000 Years of Interaction, ed. by Umberto Albarella, Keith Dobney, Anton Ervynck, and Peter Rowley-Conwy [Oxford: Oxford University Press, 2007], Figure 1.3. Reproduced with permission from Colin Groves.)

regularly imported to Britain and other European countries to improve carcass qualities of the local breeds. In this way most of the Western breeds of pig were transformed from the lean, leggy, long-faced breeds, such as the old Berkshire, into the rotund, obese, short-faced breeds such as the Middle White, which is now a rare breed.

Domestic and Feral Dogs, *Canis familiaris*, in India and East Asia

There are three species of wild canid that are locally widespread over India and southern Asia: wolf, *Canis lupus*; golden jackal, *Canis aureus*; and the red dog, or dhole, *Cuon alpinus*. Molecular studies have shown that the wolf alone is the ancestor of the dog, and the Indian wolf, a small distinctive subspecies named *Canis lupus pallipes*, is often claimed on the basis of morphology to be the ancestor of the Asian domestic dogs, particularly the ancient populations of primitive, feral dogs, known as Indian pariah dogs.[23]

In 2002 Peter Savolainen and colleagues published genetic evidence that suggested an East Asian origin for all domestic dogs, which they dated to around 15,000 years ago.[24] From the results of a further study based on the mtDNA data from a large sample of dogs from across the Old World, the team of geneticists narrowed the place of origin for all domestic dogs to the single region of China south of the Chang (Yangtze) River. They postulated that the earliest domestication of all dogs occurred less than 16,300 years ago from several hundred wolves, which were most probably of the small Chinese race *Canis lupus chanco* (see Chapter 1 for alternative archaeozoological evidence for the Eurasian origin of domestic dogs). The authors for this presumed Chinese origin wrote that "the place and time coincide approximately with

Figure 46. This dog could be a descendant of the earliest tamed wolves from south China. (Reproduced with permission from Ya-Ping Zhang.)

the origin of rice agriculture, suggesting that the dogs may have originated among sedentary hunter-gatherers or early farmers, and the numerous founders indicate that wolf taming was an important culture trait."[25]

The eating of dog meat is an ancient custom in China, which, as quoted by Epstein,[26] can be traced back to the sacrificial consumption of the flesh of wolves, and this could have been the impetus for their first taming and domestication. Carnivore meat is still relished in China, and it has been calculated that in Asia, about thirteen to sixteen million dogs and four million cats are eaten each year.[27]

The history and breeds of dogs in China and Tibet have been described in considerable detail by Epstein, who divided them into four groups: pariah; shepherd-dogs, including mastiffs; greyhounds; and toy dogs. The best known of the Chinese breeds in the West are the black-tongued Chow Chow, which was "improved" in Britain from imported pariah dogs, and the toy dogs, especially the Pekingese. Epstein stated that "the use of the term 'pai' for a short-legged, short-headed type of dog, 'which belongs under the table,' at the end of the first century A.D. provides certain evidence that at this time toy dogs were well established in China."[28] Pekingese dogs were first seen in the West in 1860 when a few were brought to England after the sacking of the Summer Palace in Beijing (Peking). According to the history of dogs by E. C. Ash, Pekingese were rarely seen in China, as they were protected by the Imperial family in the Holy City.[29]

There were two main types of dog traditionally bred in Tibet, and both have become popular breeds in the West: first, the Tibetan mastiff, which according to Epstein was first recorded in China in 1121 BCE,[30] while Marco Polo described Tibetan mastiffs "as big as donkeys, very good at pulling down game."[31] And second, the Lhasa apso and the Tibetan terrier, described by Epstein as small dogs that were bred in monasteries in the interior of Tibet and sold to the nomad tribes as highly prized mascots. The Lhasa apso is known in Tibet as "*abso seng kye*," the "bark sentinel lion dog."[32]

In Japan, the remains of dogs are frequently retrieved from the earliest prehistoric sites, known as the Jomon period, which stretched from 13000 to 300 BCE. The Japanese islands, being isolated from the mainland of Asia, have provided a valuable resource for molecular investigation into the origin and ancestry of these ancient dogs. Naotaka Ishiguro and colleagues were able to obtain ancient mtDNA from 120 dog bones excavated from twenty-seven archaeological sites dating from the earliest Jomon period to the Okhotsk period, which is dated from the seventh to the twelfth centuries CE. The results of this study indicate that there were two main lineages of Japanese dogs: an ancient population brought from Southeast Asia in the Jomon period and a later migration from the Korean peninsula.[33]

As in China and Tibet, there has been a long tradition of breeding toy dogs in Japan, and Ash described the recent history of the breeds, as well as giving a reference to the importation of British dogs in 1614 CE: "a masttife, a watter spaniel and a faire grayhound."[34]

Domestic Cats in East Asia

It has long been open to question as to whether the Burmese and Siamese breeds of cat are descended directly from the Asian subspecies of wild cat, *Felis silvestris ornata* from Central Asia, or *Felis silvestris bieti* from China. This has seemed unlikely because more than any other mammal, apart perhaps from the commensal rats and mice, domestic cats have moved around the world traveling with people in ships. Also, until modern times, reproduction in cats has generally not been under human control, and there has been no artificial selection for distinctive breeds, so it has been assumed that worldwide populations of cats have freely interbred with no breeds restricted to separate countries.

Advances in molecular analysis now indicate with a fair degree of certainty that all cats throughout the world were originally descended from the Near Eastern wild cat, *Felis silvestris lybica*.[35] However, the molecular investigations of Monika Lipinski and colleagues have verified that the Southeast Asian breeds, including Birman, Burmese, Havana Brown, Korat, Siamese, and Singapura, do form a grouping that is distinct and at the opposite end of the genetic spectrum from the Western breeds. This indicates that these Southeast Asian populations of cats have been relatively isolated from other regional populations for a long time and that the breeds were kept separate from each other.[36]

Domestic Fowl, *Gallus domesticus*

Unlike the species of domestic cattle that originated in Asia, which have never spread widely throughout the rest of the world, the domestic fowl or chickens of Asia are everywhere. With a worldwide distribution, there are probably more chickens living than any other species of domestic animal, and they provide the most efficient and cheapest form of protein, for which the suffering to the birds is sadly hardly ever taken into account.

Domestic chickens are all descended from the jungle fowl of the hot regions of Southeast Asia, of which four species are known. There used to be argument about which of these species was the predominant ancestor of the domestic chicken, with most authors agreeing that the red jungle fowl, *Gallus gallus*, was the most likely. This species occurs wild in northeastern India, southern China, and south to Sumatra and Java.[37] However, a recent genetic study has shown that the yellow skin of domestic chickens is derived not from the red jungle fowl but from the gray jungle fowl, *Gallus sonneratii*, which is found today as a wild bird only in India.[38] Modern domestic chickens are thereby proved to be hybrids of the two species.

Mohenjo-Daro used to be cited as the earliest location for the domestic chicken, dated to 2500–2100 BCE, but using the evidence of bone remains from archaeological sites, Barbara West and Ben-Xiong Zhou concluded that domestic chickens had been taken north to China by 6000 BCE from Southeast Asia, where they must have been domesticated well before that date.[39]

Domestic chickens slowly moved west through Babylon, reaching Egypt in the Ptolemaic period as recorded by Diodorus Siculus and mentioned in Chapter 4. By the time of Cato (234–149 BCE), who was the earliest notable writer on agriculture, chickens had reached Italy, but Cato only describes how they should be fattened, like geese, and they were probably rather recent imports.[40] However, Varro (116–17 BCE), the next adviser on farming, was completely familiar with "barnyard fowls," of which there were already several breeds, and he gives detailed instructions on breeding, setting eggs under broody hens, and fattening.[41] Like Cato, Varro does not mention eggs as food for humans, and neither does Columella (c. 4–70 CE), who with his usual great attention to detail wrote at extensive length on the housing, care, and feeding of chickens.[42]

The great success and wide distribution of the jungle fowl as a domesticate, historically, must be due to its versatile feeding habits, but even more to the territorial nesting habit of hens and their inability to fly over any great distance. Guinea fowl, which were also bred for eating in the ancient world, would not have been so successful as domesticates because they fly well, as do pheasants, partridges, and other "game" birds.

Apart from providing meat, eggs, and feathers, the other universal function of the domestic fowl has been to provide entertainment in the "sport" of cockfighting, which is first illustrated in seals from Mohenjo-Daro. The "sport" later spread throughout the ancient world and was popular in Classical Greece. However, Columella disapproved of it, as much as most people today who are concerned with animal welfare, and he wrote: "We take most pleasure in our own native breed; however, we lack the zeal displayed by the Greeks who prepared the fiercest birds they could find for contests and fighting. Our aim is to establish a source of income for an industrious master of a house, not for a trainer of quarrelsome birds, whose whole patrimony, pledged in a gamble, generally is snatched away from him by a victorious fighting-cock."[43]

Indian Peafowl, *Pavo cristatus*, and Green Peafowl, *Pavo muticus*

Indian peafowl occur wild throughout the Indian subcontinent, while green peafowl are found in small numbers in southern China and south to the islands of Java. In the Indian peafowl, the peacock has the magnificent, well-known fan-shaped tail of iridescent feathers, while the peahen is browner in color and has a much shorter tail. The two sexes of the green peafowl are much closer to each other in appearance, both having bright green plumage.

Indian peafowl have an ancient history as domesticates, and according to Iain Grahame, the birds have been famous in art, in legend, and in literature for over 3,000 years.[44] In the Old Testament account of Solomon's import of precious objects and animals there is perhaps the first mention of peacocks in the Near East: "For the King had at sea a navy of Tharshish with the navy of Hiram: once in three years came the navy of Tharshish, bringing gold, and silver, ivory, and apes, and peacocks."[45] The date of Solomon is uncertain, but if he lived in the first century BCE, then his peacocks could have been at about the same time as the mention of peacocks in Babylon by the Greek historian Diodorus Siculus, who lived from 80 BCE to 20 CE: "Babylonia, for instance, produces a multitude of peacocks which have blossomed out with colours of every kind."[46]

In ancient Greece, the peacock became sacred to Hera, the queen of heaven, because of the stars on its tail. Later, the Romans bred peafowl in large numbers for eating, and there are many illustrations of peacocks in their mosaics and wall paintings. Cato (234–149 BCE) does not mention peacocks in his writings on agriculture, so it seems the birds did not reach Italy until the time of the next Roman agriculturalist, Varro, who wrote, "As to peafowl, it is within our memory that flocks of them began to be kept and sold at a high price."[47] Not long after this date, as recorded by the most prolific writer on Roman agriculture, Columella, the care and breeding of peafowl were well understood, but "it calls for the attention of the city-dwelling householder rather than of the surly countryman."[48]

Goldfish, *Carassius auratus*

All ancient and modern goldfish are descended from wild carp that live mostly in China. The wild fish have olive green or brown backs, golden or silver sides, and silver undersides. There are two species of these wild carp: *Carassius auratus auratus*, the gibel carp, and *Carassius auratus carassius*, the crucian carp. Zhang Zhong-ge, an authority on the history of goldfish, believed that the gibel carp was the most likely progenitor of the domestic goldfish.[49] This has now been supported in a molecular study by Tomoyoshi Komiyama and colleagues, and it seems probable that all goldfish are descended from the Chinese gibel carp, including those from the long tradition of breeding exotic shapes and colors of goldfish in Japan.[50]

Goldfish were first bred under domestication during the Sung dynasty (960–1278 CE) in

Figure 47. Exotic goldfish in an aquarium. (Source: Bernd Brunner, *The Ocean at Home: An Illustrated History of the Aquarium* [New York: Princeton Architectural Press, 2005], 79. Reproduced with permission.)

two fish ponds: one was below Yuebo (Moonlit Waves) House in Xiushan County, and the other was Hangzhou, where goldfish were found in the ravines below Luohe (Six Harmonies) Pagoda. Later, goldfish were kept in isolated ponds within house compounds, where the fish were individually fed and artificially selected so that new breeds were produced.[51]

Goldfish were first introduced to Japan c. 1500 CE and to Europe from the beginning of the seventeenth century.[52] It may have been an imported, rare goldfish that Samuel Pepys saw on 28 May 1665 and described in his diary: "Thence to see my Lady Pen, where my wife and I were shown a fine rarity: of fishes kept in a glass of water, that will live so for ever; and finely marked they are, being foreign."

Domesticates in Oceania

Oceania is a geographical region that includes the continents of Austra-lia and New Zealand as well as Papua New Guinea and a very large number of smaller islands across the Pacific Ocean, as shown figure 48.

With acceptance of the theory that the first anatomically modern humans moved out of Africa about 200,000 years ago, most anthropologists now agree that these first humans then traveled through Asia and India, keeping to the coasts, until they reached East Asia about 70,000 years ago. Sea levels were lower at this time, which is termed the Lower Pleniglacial, when the north was covered in thick ice sheets and there were land bridges between Southeast Asia, New Guinea, Australia, and Tasmania. It is not surprising, therefore, that the earliest evidence of these first humans has been found in Australia dating to around 55,000 years ago, many thousands of years before the remains of anatomically modern humans have been found in Europe.

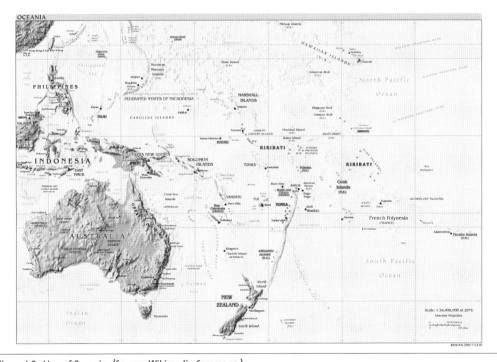

Figure 48. Map of Oceania. (Source: Wikimedia Commons.)

As Manfred Kayser has pointed out in a recent review, human migration into the lands of Oceania is unique in that it encompasses evidence for both the first out-of-Africa expansion of anatomically modern humans to New Guinea and Australia as well as the latest occupation of some Polynesian islands. New Guinea has other anthropological peculiarities, which include the 1,000 often very distinct languages, the independent and early development of agriculture in the highlands about 10,000 years ago, and the long-term isolation of the interior of New Guinea, which lasted until the 1930s.[1]

There is also the intriguing problem of the origin of the tiny hominins whose remains have been found on the Indonesian island of Flores, where they lived with pygmy elephants (stegodon) and komodo dragons from about 95,000 to 13,000 years ago. These miniature persons, only three feet tall, may be a separate species of human and have been given the provisional name of *Homo floresiensis*, although their small size may just be the result of island dwarfing, as with the fossil elephants.[2]

Between 10,000 and 8,000 years ago, as in the north with the melting of the ice and worldwide rising sea levels, lands in the Pacific became separated as islands from each other. Australia was cut off from New Guinea, as was Tasmania from the south of Australia, but by this time humans had learned how to travel by boat, taking live animals with them.

An archaeozoolgical peculiarity of ancient Oceania is that the only domestic animals that were ever taken to the islands, including New Guinea and Australia, were dogs, pigs, and chickens, with rats as commensals. Susan Bulmer has cited evidence for pig remains dated to 8,000 years ago on the island of New Ireland in the Bismarck Archipelago, but no direct evidence for dogs this early, although it is hard to believe they did not reach the islands at the same time as pigs.[3] It is curious, however, that there is no evidence that the people who took the first dogs to Australia also took pigs. Perhaps they did, but perhaps as hunter-gatherers and fisherfolk they found no use for pigs, and the arid climate inland prevented pigs from reproducing as populations in the wild. The dogs, on the other hand, became such successful wild carnivores that there is a case for classifying them as *Canis dingo*, a separate species from the domestic dog, *Canis familiaris*.

DOGS IN ANCIENT OCEANIA

New Guinea Singing Dog

The first island of Oceania to which dogs were taken may have been New Guinea, and together with pigs, their function could have been to help with hunting, but also to provide meat, hides, and fur in a land that was dominated by wild birds: the beautiful and unique birds of paradise. It may be assumed that the dogs spread rapidly into the forests and bred as wild populations. Subjected to natural selection in the high-altitude mountains and forest environment, they evolved into a distinct morphological type with the tawny-yellow coat of the dingo and the singing howl, which carried over great distances and held the packs together.

The earliest positive ancient evidence for dogs in the New Guinea Highlands comes from hunting shelters dated 6,000–5,000 years ago.[4] The survival of this ancient breed of dogs in the New Guinea Highlands, which appeared to belong to a separate race from the common domestic dogs of Asia, remained unknown until a skull and skin were taken to the Queensland

Museum in Australia in about 1918. The dog had been obtained from local people at 7,000 feet altitude on Mount Scratchley. The first pair of live "mountain" dogs was brought out of the Southern Highlands in 1956, presented to Sir Edward Hallstrom, and donated by him to Taronga Zoo in Sydney, Australia. These dogs were named as a distinct species, the New Guinea singing dog, *Canis hallstromi*, by Ellis Troughton in 1957.[5] Today, descendants of these dogs are sometimes held in other zoos and are kept as a rare companion breed, but they are not easy to train. The morphology and behavior of New Guinea singing dogs have been described by Janice Koler-Matznick and colleagues.[6]

Australian Dingo

Unlike the New Guinea singing dogs, which only became known to outsiders at the beginning of the twentieth century, the dingo, the wild dog of Australia, has been known since the seventeenth century and has been persecuted as vermin and an unwanted predator on introduced domestic livestock, especially sheep, for much of that time. All that is known of when dingoes arrived in Australia is that they are descended from domestic dogs that were taken to the continent by boat relatively late in the timescale of immigration by the Aboriginal people, who may have been there for the past 55,000 years. The earliest positive evidence for dingoes, on the other hand, is from a number of archaeological sites dated as late as 5,000 to 3,500 years ago. They certainly could not have been introduced before 10,000 years ago because no remains of dogs have ever been found in archaeological sites in Tasmania, which was cut off from Australia by the Bass Straits at this time.

During the twentieth century many biologists and archaeozoologists studied the skeletal anatomy of the dingo in attempts to discover its origins and relationships with other primitive breeds of dog. All that could be determined was that the dingo was morphologically similar to these other breeds, which, for the dogs in Thailand, L. K. Corbett named "Thai dingoes."[7] With the advance of molecular biology, however, it has now been possible to elucidate the genetic relationships of these dogs and thereby their most probable origins. This has been carried out by Peter Savolainen, who analyzed 211 Australian dingoes as well as a large sample of dogs and wolves from worldwide sources and nineteen pre-European archaeological dog samples from Polynesia. Savolainen and colleagues found that a majority of the dingoes had mtDNA type A29, which was only found in dogs from East Asia and Arctic America, whereas eighteen of the nineteen other types of mtDNA were unique to dingoes. The mean genetic distance to A29 among the dingo mtDNA sequences indicated an origin 5,000 years ago. From these results it was deduced that dingoes have an origin from domesticated dogs coming from East Asia. They were introduced from a small number of dogs, possibly at a single occasion, and have since lived isolated from other dog populations.[8] This result is, of course, not unexpected and supports the assumptions from all the countless measurements that have been taken of dingo, wolf, and dog skulls over the past fifty years.

Apart from its genetic descent from East Asian dogs, the uniform morphology of the dingo—with its tawny-yellow coat, single annual breeding cycle, and distinctive social behavioral patterns—links these dogs with primitive and feral dogs in many parts of the world as well as with the New Guinea singing dog. However, like the singing dog, the dingo's isolation from interbreeding with European dogs, until recently, kept its genotype separate and justified its name of *Canis dingo*. This means that dingoes as a breed are a valuable living relic of early domestic dogs, and they are also part of the living heritage of hunter-gatherer culture, but the purebred dingo is becoming a rare animal that should be conserved, for its final extinction would be a loss on many counts.

The complicated role played by dingoes in the lives of the indigenous Australians must mirror that of many prehistoric communities of hunter-gatherers with their dogs, which were scavengers, hunting companions, bed warmers, religious symbols, and very occasionally providers of meat and bodily parts. However, relatively few original observations have been published of the association between the indigenous Australians and the dingo. An early account was by C. Lumholtz in 1889, followed by the reviews of M. J. Meggitt in 1965 and Bradley Smith and Carla Litchfield in 2009.[9]

DOGS IN THE ANCIENT PACIFIC ISLANDS

The role of dogs in the different islands of the wider Pacific was reviewed in valuable detail by Margaret Titcomb in 1969, who introduced her account with the comment:

> It is likely that all Pacific islanders loaded their sea-going craft with food for the voyage, whether a deep-sea fishing voyage, or a planned trip to another island. Ideally for a voyage of any length, the dog, the pig, and the chicken were carried along. On most islands they survived. . . . Most dogs did not live very long. Puppies were more delicious than older dogs. Only those used for breeding could have survived puppyhood. . . . The Pacific dog never developed a life of its own, never became feral, for food in the forest was lacking.[10]

As well as providing much-needed meat, dogs were of great religious significance on many of the islands, and their teeth, bones, and hair were used in rituals of many different kinds as well as in clothing: "The Tahitians went to battle in their best clothes. . . . On the breast they wore a handsome military gorget wrought with mother-of-pearl shells, feathers, dog's hair, white and coloured."[11]

PIGS AND CHICKENS IN ANCIENT OCEANIA

The several species of wild pigs mixed with domestic breeds and hybrids of wild and domestic pigs on many of the Oceanic islands means that distinguishing between them and sorting out their history is a formidable task. There are up to five indigenous species of wild pigs in the region that have been tamed or domesticated, and these have been hybridized with the imported "normal" domestic pigs, descended from wild boar (*Sus scrofa*). Colin Groves, in his *Ancestors for the Pigs*, wrote a comprehensive description of all the species in 1981, based on their skeletal morphology and distribution,[12] and he followed this with an up-to-date summary in 2007 (see figure 45).[13] Groves separates the Asian pigs into two overall groups, the *scrofa* group and the *verrucosus* group, the latter of which includes the wild Oceanic pigs, these being three species of warty pigs, the bearded pig, the Palawan pig, the Visayan pig, and the Philippine pig.

As with so many other topics in archaeozoology, new research in molecular biology has provided the most detailed account of the ancient spread of domestic pigs on the Oceanic islands. This can be summarized in the words of Greger Larson and colleagues: "Through the use of mtDNA from 781 modern and ancient *Sus* specimens, we provide evidence for an early human mediated translocation of the Sulawesi warty pig (*Sus celebensis*) to Flores and Timor and two later separate human mediated dispersals of domestic pig (*Sus scrofa*) through Island Southeast Asia into Oceania."[14] Larson and colleagues concluded that domestic pigs of the *Sus scrofa* clade were introduced from Island Southeast Asia to the Oceanic islands, including New Guinea, where feral populations have become separated from the domestic and are hunted as wild animals by the New Guinea Highlanders. These pigs arrived in New Guinea and the Moluccas as late as 3,500 years ago and were probably linked with the arrival of nonindigenous agriculturalists. The other later pig dispersal linked mainland East Asian pigs to western Micronesia, Taiwan, and the Philippines.

Chickens provide essential meat, as well as feathers for clothing and bedding, and it is probable that chickens were originally introduced to Island Southeast Asia and the Pacific islands at the same time as pigs and from the same geographic sources. In a study of the mtDNA of modern chickens, Y. P. Liu and colleagues identified Southeast Asia as a center for domestication of chickens,[15] and Larson and colleagues postulate that pigs and chickens were dispersed to Oceania together.[16] Liu and colleagues found nine highly divergent clades in their study, and one of these (clade D) was closely related to fowls bred for cockfighting, which has also probably been common in Oceania since chickens were first introduced to the islands.

Figure 49. New Guinea: a sow with her piglets resting in the heat of the day. (Copyright Rebecca Jewell, reproduced with permission.)

Figure 50. Solomon Islands pig on a beach at Santa Cruz. (Copyright Rebecca Jewell, reproduced with permission.)

Figure 51. A Mendana cock on Santa Cruz, Solomon Islands. (Copyright Rebecca Jewell, reproduced with permission.)

NEW GUINEA BIRDS OF PARADISE AND MAORI FEATHER CLOAKS

Birds of paradise have been admired, hunted, and traded by the indigenous people of New Guinea for thousands of years, and their feathers are used for headdresses and are important as clan totems, as symbols of status and wealth, and as a bride price. Rebecca Jewell has

Figure 52. Birds of paradise. (Source: Wikimedia Commons.)

reviewed the relationship between the tribesmen and the birds of paradise and has written of the great significance and value of the birds in the people's culture. The New Guinea Highlanders' reverence and respect for the birds has meant that these birds have been sustainably harvested for their feathers, despite the very large number that have been killed over time.

The particular trees where the birds of paradise display are preserved by the tribesmen, and their locations are handed down through the generations. It is thought that the reason why extensive exploitation of the birds over time has not resulted in extinction of the species is that the adult birds are able to breed before they are fully plumed, so they are only killed after the young have fledged.[17]

In New Zealand, cloaks made with feathers were not part of Maori culture at the time of Captain Cook's contact in 1769, but in the second part of the nineteenth century cloaks began to be made from the feathers of local birds and from introduced pheasants and peacocks.[18]

Human behavior is the same worldwide, for where precious metals and jewels can be mined, they have been used for thousands of years to decorate and give grandeur to religious and ceremonial activities. Where these are absent or unknown, as in the Pacific islands, the feathers of the most beautiful of all birds have been used for the same purposes and woven into the most elaborate objects and articles of clothing that are treasured and only worn on ceremonial occasions.

THE IMPACT OF EUROPEAN INTRODUCTIONS OF DOMESTICATES AND
THEIR FERAL DESCENDANTS IN AUSTRALIA AND NEW ZEALAND

From the end of the eighteenth century selected breeds of livestock were imported to New South Wales in Australia from British landowners and were successfully bred there in considerable numbers. One letter from the Reverend Samuel Marsden (1765–1838) sent from Sydney on 13 January 1805 to Sir Joseph Banks reflects the importance that was given to establishing what were considered to be the improved breeds of sheep:

> We have got the Spanish [merino], South-down & Tees-water breed already; tho' not general. I think the Leicestershire & Lincolnshire Breed would very much improve our flocks, could we obtain them. I have written to my agent to purchase me two Rams; and requested Mr. Campbell to bring them out provided Government will give Consent for them to come out of the country [England]. My attention has been very much turned to improve our Flocks, as far as we have the means of doing it; and have hitherto succeeded beyond my Expectation considering the poor original Stock we had to breed from.[19]

In another letter to Banks, Marsden wrote: "The ewes have lambed twice since my arrival. These sheep [Spanish merinos] will prove of great value to this Country. Our climate is so very fine and agreeable for them. I have sent to England in the *Admiral Gambier* from 2,000 to 5,000 lbs. of wool, of the best Quality I have. My flock consists of 3,700 at present."[20]

Merino sheep continued to thrive and adapt to the different Australian environments, and today they number well over 100 million, while over the past two centuries almost every other domestic species, as well as very many wild ones, have been purposefully introduced to Australia. Descendants of these introductions now number in the millions, like the vast plagues of feral rabbits and house mice that are so well known, but there are also feral herds of unexpected species like the Indian water buffalo (*Bubalus bubalus*) and dromedaries (*Camelus dromedarius*).[21] Almost equal numbers of introduced animals have been taken to New Zealand, where they also thrive. One of the few detailed studies to be made of a feral species in Australia (apart from the dingo) has been on the introduced pigs by C. A. Tisdell, who calculated that in the 1980s there was likely to be between eight and eleven million feral pigs living throughout Australia.[22]

There were no ungulates in either Australia or New Zealand before the arrival of European immigrants. In Australia, a great diversity of marsupials—especially kangaroos and wallabies—were grazers and browsers, but they are soft-footed, and their replacement by millions of hard-hoofed livestock has led to severe deterioration of the grasslands. The removal of vegetation from the soil over huge areas has led to severe droughts, forest fires, and increasing climate change. In New Zealand, before the arrival of the Maoris with their dogs from Polynesia, sometime before the thirteenth century CE, the niche of grazers and browsers was filled by the flightless birds known as moas, some of which were the largest birds ever to have lived on Earth. When the Dutch explorer Abel Janszoon Tasman (1603–1659) reached the west coast of South Island in 1642, the moas had been exterminated by the Maoris, and the only wild mammals on the islands were two species of bat and Polynesian rats, which the Maoris had brought with them, probably unintentionally.

The arrival of foreign immigrants over the past centuries, with their own national cultures and their grazing livestock, has transformed both Australia and New Zealand beyond recognition and created what Alfred Crosby has described as Neo-Europes.[23] The reason, of course, is because people who emigrate to new and strange lands want to make them as similar as possible to their place "back home."

Domesticates in Africa
South of the Sahara

THE FOSSIL RECORD SHOWS THAT THE FIRST HUMANS EVOLVED IN AFRICA 200,000 years ago, and from this continent they slowly traveled across the world, reaching Australia 60,000 years ago, Europe 40,000 years ago, and the Americas 15,000 years ago. Linguistic and genetic evidence indicates that the people who stayed behind are represented today by the last of the San Bushmen. The genetic structure of Bushmen individuals has shown that in their genomic divergencies, they are more different from each other than a European is from an Asian, and they belong to the oldest known lineage of modern humans.[1] These people are hunter-gatherers who live in southern Africa in small protected areas that are in marginal and hardship zones to where they have been driven over the last two millennia, first by the diffusion southward of Bantu-speaking agriculturalists and pastoralists, and later by foreign colonists.[2]

Until recent rethinking, it was generally accepted by linguists and archaeologists that the first proto-Bantu farming began in the forested region of the Cameroons around 4,000 years ago. It was presumed that population numbers had greatly increased after the people had learned how to forge iron, cultivate cereal crops, and herd livestock as nomadic pastoralists. Migration then became inevitable, and it was hypothesized that around three millennia ago, the Bantu-speaking peoples began to move south and east from West Africa, taking their cattle, sheep, and goats with them; planting their crops with their new iron tools; and building nation-states across the continent. The theory was that while cattle moved south with the eastern stream, sheep moved with the Bantu-speakers down the western side of the continent to arrive at the Cape around 500 CE. However, the remains of very few domestic livestock have been retrieved from the early Iron Age sites in Cameroon, while there is evidence of pastoralists in northern Kenya during the third millennium BCE and in southern Kenya by the end of the second millennium BCE, so the Cameroon hypothesis may have to be revised.[3]

By whatever routes they followed, these early Bantu-speaking pastoralists slowly spread everywhere south of the Sahara, but they did not enter an empty land. It was inhabited by hunter-gatherers who lived off the teeming wildlife but who soon began to take advantage of the way of life of the incoming herders. The spread of domestic animals southward through the grasslands of Africa probably had some similarities with the Neolithic spread of the same livestock species from the Near East westward across Europe, only it occurred several millennia later. People who had lived for thousands of years by hunting and gathering slowly changed to the ownership of tamed livestock, and, as in other parts of the world, this change occurred in many different ways, including barter and trade, bride-price, thieving, warfare, and the migration of the people themselves with their animals. Why this change was worthwhile is more of a problem in Africa than in Europe, where it can be surmised that in the early Neolithic, livestock herding became a new means of obtaining food and resources for human populations that were expanding and causing depletion in the hunters' prey and in food that could be foraged.

For what purposes were the herds of cattle, sheep, and goats kept and tended with such care on the African grasslands, where they had to be given regular water and enclosed in a boma (stockade) every night to protect them from carnivores? It is hard to believe, at least in the early phase of livestock herding, that the animals were important as a source of meat, as there can never have been a lack of wildlife to hunt, and the endemic diseases and parasites carried by tsetse flies and tick-borne fevers must have killed off great numbers of livestock, which, being foreign immigrants, lacked immunity. However, as they settled in different regions of the vast continent with its great variations in climate, parasitic diseases, and other deleterious selective pressures, the pastoralists and their livestock developed a great diversity of adaptive strategies that enabled them to survive and flourish. The ownership of large herds of distinctive breeds of livestock, particularly cattle, then became a symbol of wealth and status for the tribal chiefs.

Study of the diverse pastoral economies in Africa has occupied anthropologists for generations since the renowned research of Evans-Pritchard on the Nuer and their cattle in the 1940s.[4] There has been less emphasis on the study of the breeds of livestock that were, and are, the basis of these economies, but within recent decades their importance and the value of the individual adaptations of breeds to local environments are becoming more generally appreciated.

DEVELOPMENT OF THE BREEDS OF AFRICAN CATTLE

The combination of anthropological studies, archaeozoology, and molecular biology has enabled the history of cattle to be better known than for any other domestic species in Africa, but there are still many questions to be answered, especially about the development of the local breeds, which are central to the history of African pastoralism. The early herders who moved south through the Sahara probably owned humpless taurine cattle while humped zebu cattle entered the continent later through the Horn of Africa from Arabia (see figure 53).[5]

By 6,000 years ago, cattle pastoralism was well established north of the Sahara, and with the increasing aridity of the desert, people began to move south with their herds of shorthorn and longhorn cattle until they met the tsetse fly belt. Tsetse flies carry trypanosome parasites that give cattle and other artiodactyls the disease of trypanosomiasis, which can be lethal, and they give sleeping sickness to humans. One longhorn breed that is probably descended from these earliest migrants is the present-day West African N'Dama, which is renowned for its immunity to trypanosomes. This indigenous breed is similar in the size of its skeleton to the very small cattle whose bones have been identified from the site of Kintampo in Ghana, dated to around 1500 BCE.[6] At first many of the cattle that reached the tsetse fly belt in West Africa would have succumbed to the parasites, but over the centuries increasing numbers would have inherited immunity until the N'Dama breed became established.

By the time that humped cattle began to be introduced to the continent from Arabia, perhaps between the first millennium BCE and the first century CE,[7] Bantu-speaking pastoralists had probably spread with their humpless taurine cattle through tsetse-free corridors (see figure 53)

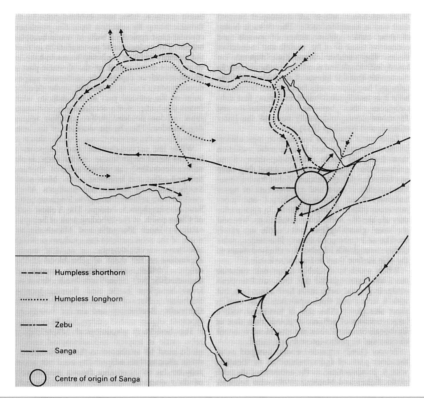

Figure 53. Geographical origins of indigenous cattle in Africa. The earliest African cattle are now considered to have been derived from wild *Bos primigenius* in North Africa, with later imports from western Asia. (Source: H. Epstein and I. L. Mason, "Cattle," in *Evolution of Domesticated Animals*, ed. by I. L. Mason [London: Longman, 1984], 13, Figure 2.1.)

over most of the grasslands of West and East Africa. In his classic work on the domestic animals of Africa, H. Epstein divided zebu cattle into thoracic-humped, with the hump over the thoracic vertebrae, and cervico-thoracic humped, with the hump over the posterior part of the neck,[8] but the distinction is not always clear, and in many African breeds the hump can be in an intermediate position. In their molecular analysis of African humped cattle, Ronan Loftus and Patrick Cunningham therefore use the term "zeboid" to describe cattle with varying degrees of zebulike features.[9] This term also includes Sanga cattle, breeds that are descended from ancient hybrids between indigenous humpless breeds and early introductions of zebu cattle. It is most likely that the two positions of humps were developed as a result of a combination of artificial and natural selection in domestic cattle that were living in hot, arid zones, in a parallel way to the development of the fat tail and fat rump in breeds of sheep living in the same environments.

At least 120 different breeds of unhumped and zeboid cattle have been developed throughout Africa, and many are of ancient origin and have truly remarkable adaptive strategies for life in environments that appear to be totally unsuitable for them. Perhaps the most remarkable of these are the Kuru cattle of Lake Chad. This is a very large, indigenous breed with hugely inflated horns; the cattle spend much of their time grazing in the swamps, and they swim from island to island in the lake.

In a detailed molecular study of living Kuru cattle, Ciaran Meghen and colleagues have shown that they are not markedly divergent from the N'Dama, but also, although they are unhumped, they do show some genetic evidence of crossbreeding with zeboid cattle.[10] Another

Figure 54. Kuru cattle of Lake Chad. (Source: Ciaran Meghen, David MacHugh, B. Sauveroche, G. Kana, and Dan Bradley, "Characterization of the Kuri Cattle of Lake Chad Using Molecular Genetic Techniques," in *The Origins and Development of African Livestock: Genetics, Linguistics and Ethnography*, ed. by Roger M. Blench and Kevin C. MacDonald [London: UCL Press, 2000], 260, Figure 15.1. Reproduced with permission.)

distinctive West African breed is the White Fulani of the pastoralists living across the Sahel from Senegal to Lake Chad. They are Sanga cattle with lyre-shaped horns, indicating that they probably originated as longhorned migrants from the north, which were later interbred with zeboid cattle. Many of the Fulani people now live in settled communities and sell their cattle for beef, but there are still pastoralists who herd their Fulani cattle in the old ways, use them for milk and draft, but do not slaughter them, and live in highly organized transhumant camps with strict rules and rituals.[11]

In East Africa, Ankole Sanga cattle are the most distinctive, especially the Ankole of the Watusi, or Tutsi, warriors, which have excessively long horns. The Watusi are the tallest and most fine-boned people in the world, and their cattle match them for elegance. According to Valerie Porter, the sacred cows that used to belong only to the kings of Ruanda (Inyambo) have the longest horns of all, with a span of 240 centimeters.[12]

The best-known pastoralists of East Africa are the Maasai, the seminomadic tribe of specialized cattle herders who have traditionally inhabited arid grasslands with an unreliable rainfall. The Maasai management of their cattle and their lineages have been studied in Kenya by Kathleen Ryan and colleagues, focusing on how cattle are acquired, how they are exchanged, and how and under what circumstances they die. Ryan and colleagues conclude that it is through the genealogies and the remembered transactions in their cattle that past and present social relations in the human communities are preserved.[13] An earlier study by John Galaty also looked at how the Maasai obtained their cattle, but he recorded their enormously detailed methods of classifying and naming individual animals and lineages. The Maasai do not own a particular breed of cattle but have traditionally obtained their herds through barter and raiding from other tribes.

Cattle are the core and the mainstay of the Maasai, as they are of other pastoralist tribes in Africa. Apart from their function in society, Maasai cattle supply meat, milk, and blood for food. As milk is such an important part of the diet of cattle herders in Africa, geneticists have attempted to analyze what proportion of the adults are lactose tolerant—that is, whether they retain the enzyme lactase into adult life, which enables them to digest milk. For the Maasai, the results of testing children and adults for lactase persistence has not been definitive, but in general Sarah Tishkoff and colleagues have found genetic adaptation for the digestion of milk in a proportion of the herders, which they maintain shows convergent evolution with the genetic ability to digest milk as adults in Europeans.[14]

The spread of cattle southward from East Africa, which began around 2,000 years ago, continued until it reached the Cape, with many distinctive Sanga breeds being developed in adaptation to the local conditions of the different regions. Where land was fertile, the Bantu-speaking pastoralists settled and built houses and religious centers out of dry stone walling, and became nations. The greatest of these settlements was Great Zimbabwe, which became the center of the Shona nation and reached its zenith between the thirteenth and fifteenth centuries CE. During this time it was a large town covering some 78 hectares and housing an estimated 18,000 people. Great Zimbabwe was the center of a network of sites that flourished with the export of gold to the Indian Ocean. However, cattle herding remained the basis of the economy, and one suggestion for the apparent rather sudden collapse and abandonment of the town is that the surrounding grasslands became overgrazed and could no longer support the livestock that were essential for the survival of the people.[15] It was probably the medieval Shona inhabitants of Zimbabwe who developed the Mashona breed of cattle, which until the recent political turmoil was widespread in east Zimbabwe.

In South Africa, there were two main groups of inhabitants before the southward expansion of the Bantu-speaking people. They were the indigenous Khoi people who belonged to a single linguistic group but were divided into the Khoisan (Bushmen), who were hunter-gatherers and who survive today in very small numbers, and the Khoikhoi (Hottentots), who were pastoralists who herded cattle and sheep but who were almost all exterminated by European colonists in the nineteenth century.

The best-known original breed of cattle in South Africa is the Afrikander Sanga, which was first bred by the Khoikhoi pastoralists from cattle that reached the Cape around 1,900 years ago, at least 7,000 kilometers from where they were first domesticated.[16] Throughout the south, these cattle were not only bred to provide food, but they had the very important added use as riding and draft oxen.

Figure 55. A khoi family on the move, 1834. (Source: Painting by Charles Davidson Bell, no. 14 in his *Scraps From My South African Sketch Books* compiled in 1848. Reproduced with permission from the Africana Library/Museum, Johannesburg.)

PIGS, SHEEP, AND GOATS IN AFRICA SOUTH OF THE SAHARA

Three species of wild pig are endemic to Africa south of the Sahara: the widespread warthog, *Phacochoerus aethiopicus*; the bush pig, *Potamochoerus porcus*, which is also widespread and found in forests and grasslands; and the giant forest hog, *Hylochoerus meinertzhagerni*, found in the central African forest belt. However, there are little or no archaeozoological records of domestic pigs, *Sus domesticus*, and they were clearly not bred by the indigenous people until introduced by Europeans.

For the past two millennia there have been four different groups of sheep in Africa: hair sheep (that is, with no woolly fleece), thin-tailed wool sheep, fat-tailed sheep, and fat-rumped sheep, all descended from the wild Asiatic mouflon, *Ovis orientalis*. These sheep derive from introduced stock that originally came from western Asia, and they probably reached West and East Africa south of the Sahara with pastoralists by the same routes as the cattle.

The most commonly held view about how and when sheep spread farther south, eventually reaching the Cape, is that proposed by Richard Klein based on the identification of domestic livestock remains from Iron Age sites. Klein suggested that sheep were introduced along the west coast of South Africa around 2,000 years ago, while cattle moved south along the east coast.[17] However, this is only a hypothesis, and two other models have been suggested: East African herders with pottery, stone bowls, and sheep moved south until they met the San hunters of southern Africa, who already had domestic animals; and Khoisan groups living in East Africa obtained livestock from local people and moved south with them. These two models would explain the presence of sheep in the Cape 200 years before the arrival of Bantu-speakers, for which there is archaeological evidence, and it is also suggested by the Bushmen (Khoisan) rock art of fat-tailed sheep and cattle.[18] The fat tail of the sheep was an extremely valuable resource of fat to pastoralists throughout the continent, as it was to the Dutch settlers who arrived at the Cape in 1652.

Until recent times it appears that goats have been less important than sheep everywhere in Africa south of the Sahara, but as sheep and goat remains have not been distinguished on many archaeological sites, this may be a false assumption. There is no doubt, however, of the high value of the fat tail and fat rump of African sheep, which would make them preferred over goats, and also there are no recorded Bushmen paintings of goats.

The earliest remains of undifferentiated sheep/goats in East Africa have been recorded from third millennium BCE sites on the northern border of Kenya with Ethiopia (the Lake Turkana basin). They can then be traced slowly southward through the first millennium BCE site of Prolonged Drift and then to early Iron Age sites north and south of the Zambesi River.[19]

Like sheep, all African goats are immigrants from western Asia, where their progenitor was the wild scimitar horned goat, *Capra aegagrus*. The remains of dwarf goats have been identified from the second millennium BCE sites of Ntereso and Kintampo in Ghana, along with the remains of small cattle, and, like the small N'Dama cattle, the dwarf goats of West Africa today are resistant to the trypanosome parasites carried by tsetse flies.[20] A few bones of goats have been identified from Iron Age sites in South Africa, but nowhere in similar numbers to those of sheep.

In conclusion, when reviewing the history of domestic livestock in Africa south of the Sahara, it is important to emphasize that as great a diversity of breeds has evolved in the past 2,000 years in adaptation to the many different environments of this vast continent as in all of Eurasia. The tragedy is that most of these breeds, or landraces,[21] have been ignored by immigrant livestock farmers. They have been disparaged under the general term "native breeds" and have been allowed to die out or have been crossbred with European breeds in attempts at improvement. Where a few breeds have survived, such as the dwarf trypanosome-tolerant cattle and goats of West Africa, it has been finally realized that their ancient lineage of genomic diversity is of unique value, and, if lost, it can never be recovered for future breeding programs.

DOMESTIC FOWLS

Although there are several species of wild Guinea fowl that are widespread in African savannas and forests, there is no archaeozoological evidence that this endemic bird was ever bred in captivity before colonial times. The early remains of imported domestic chickens (*Gallus gallus*), on the other hand, have been identified in West Africa at the site of Jene-jeno in Mali dated to around 500–800 CE; in East Africa, from two Iron Age sites in Mozambique; and in Natal, from the Iron Age site of Ndondondwane.[22]

ONE-HUMPED CAMEL

The one-humped camel, or dromedary (*Camelus dromedarius*), is the domestic animal over all others that enables humans to live in some of the most inhospitable regions of the world. There are no wild dromedaries in existence, and it is not known when they died out or exactly where or when the first wild dromedaries were domesticated, although it is assumed from archaeological evidence that it was in Arabia. From there camels entered Egypt, probably during the third millennium BCE, and they must also have been brought into Africa farther south across the Horn of Africa, where there are large numbers today.

Camels are currently found in all the countries of North Africa and the Sahara. South of the desert, camels are still herded by nomads across the countries of the Sahel, from Senegal in the west to Somalia and the north of Kenya in the east.

Camels are both grazers and browsers, and they can survive well in desert habitats because they keep on moving while grazing. They may cover 5 kilometers in two and a half hours, so they do not degrade the desert vegetation, unlike more slow-moving cattle and goats. In their comprehensive book on camels Hilde Gauthier-Pilter and Anne Innis Dagg describe how camels thrive on hard, dry, thorny plants, and "no matter how rich or how poor the quality of the vegetation, they take only a few bites from any one plant before moving to another."[23] They can go without water for longer than any other domestic mammal, but when camels are driven across the desert, where there is no surface water, they are dependent on their herders for water drawn in buckets from wells, some of which are ancient in origin. Probably for 1,000 years or more, by supplying their camels with water, but allowing them to graze freely, desert nomads have exploited a huge region across Africa that cannot be used for traditional agriculture. From the camels the nomads get milk, transport, meat, hair, leather, and dung for fuel.

DONKEYS, HORSES, AND MULES

Until the European incursions of southern Africa, the majority of indigenous pastoralists and farmers had no need of animals for transport, but where status required it for a chief to be carried in state, the ox usually took the place of the horse—the reason being, as colonists found to their cost, imported horses were very often struck down by trypanosomes from tsetse flies and by the highly infectious and deadly disease of African horse-sickness. This is caused by a virus of the genus *Orbivirus*, which is spread by biting insects, and where it gets a hold the virus leads to nearly 90 percent mortality in horses, 50 percent in mules, and 10 percent in donkeys. Imported horses and mules have little resistance to this endemic disease because their progenitor, the wild horse of Eurasia, never occurred in any part of Africa, while donkeys have considerable immunity because they are descended from the African wild ass (*Equus africanus*). However, although donkeys have been of considerable importance as pack animals since the precolonial period, their remains have seldom been found on archaeological sites south of the Sahara, so it is not possible to assess their numbers or distribution.

There are many accounts of the sudden death of teams of horses being used by Europeans to travel through southern Africa in the nineteenth century. One example is that of Lord Randolph Churchill on an expedition to Mashonaland (Zimbabwe) in 1891:

> The Bechuanaland Border Police have been losing, and are still losing, from 80 to 90 per cent of their horses. The losses of the Chartered Company have been on a similar scale, and have been equalled by those of private persons. The roadside from Tuli hither is literally strewed with dead bodies of horses and mules. . . . Mr Alfred Beit lost more than half his mules, and on reaching Fort Victoria [Masvingo] had to resort to oxen to draw his carriage and light wagons.[24]

Figure 56. Lord Randolph Churchill's mixed team of mules and oxen traveling north through sand from Fort Victoria at two and a half miles per hour. (Source: Randolph S. Churchill, *Men, Mines and Animals in South Africa* [London: Sampson Low, Marston, 1891], figure facing p. 197.)

Although the horse was little known in southern Africa, there is evidence from bone remains and a bronze hilt in the form of a man on horseback that ponies were present in Nigeria from around 1,000 years ago. Roger Blench, in a review of horses in precolonial West Africa, discussed how they were present in considerable numbers across the savannas of West Africa and that they were dwarfed to sometimes as small as a meter high at the withers. As with the dwarf cattle and goats, this was most probably a response to the harsh environment, and also, like the bovids, these ponies had evolved an immunity to trypanosomes. In his review Blench also discussed the tradition that precolonial states, such as Gonja in the north of Ghana, were founded by horsemen in the sixteenth and seventeenth centuries CE.[25]

DOMESTIC DOGS AND THE AFRICANIS

Like all other domesticates, the dog arrived late in southern Africa, but its independent way of life and rapid rate of reproduction must have meant that once a few had been introduced and were owned by Bantu-speakers and Khoikhoi pastoralists, they spread very rapidly and were probably ubiquitous by the beginning of the second millennium CE. But these dogs must have lived in permanent danger of an early death from other carnivores and baboons, which today kill many dogs. Also, as today, the dogs would have been susceptible to the many diseases that afflict canids, from rabies to tick fever.

Because dogs overlap in size with the endemic African jackals, it is not easy to distinguish their skeletal remains on archaeological sites. However, the spread of dogs southward with livestock herders can be shown by the few finds of their remains from the north-to-south sequence of archaeological sites: Esh Shaheinab and Kerma in Sudan, c. 3300 BCE–2000 BCE; Ntusi in Uganda, 895–1025 CE; Iron Age Kalomo in Zambia, 950–1000 CE; south of the Limpopo River (central southern Africa), after 600 CE.[26] As with the spread southward of sheep and cattle, there is more than one hypothesis for the arrival of dogs in the south: either after reaching South Africa the Bantu-speakers were in contact with the indigenous Khoisan peoples who obtained dogs and livestock from them by barter, trade, or thieving; or, as described above, the Khoikhoi (Hottentots), who were pastoralists with livestock and dogs, moved south from East Africa before the migration of the Bantu-speakers. The Khoikhoi would have met the Khoisan, who were hunter-gatherers and who were the indigenous people living in South Africa. To divide the Khoi peoples into two groups is really an inexcusable simplification, as they had many separate languages that were mutually unintelligible, but all that can be said here is that dogs appear to have been more popular with the pastoralists than with the hunters. In the wonderful rock art of southern Africa, which was the work of painters living hundreds of years ago, there are rather few images of dogs, and these are usually shown following people rather than in a hunt.

By the time of the first Dutch settlers in the Cape of Good Hope in 1652, dogs were everywhere, and they soon became a source of cultural division between the European purebred immigrant dogs (valuable) and the native dogs (without value). From the colonists' point of view, indigenous dogs were, and often still are, considered to be disease-ridden curs that bark incessantly and need to be controlled if not exterminated. They were commonly called by Europeans kaffir dogs or stray dogs, but they are not strays, and however mangy and starved they may look, almost all of them have owners, and many are treated with affection and get as much care

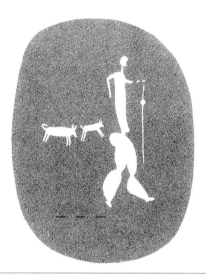

Figure 57. A woman followed by two dogs. Drakensberg Bushman painting, Site W 21. (Source: Patricia Vinnicombe, *People of the Eland: Rock Paintings of the Drakensberg Bushmen as a Reflection of Their Life and Thought* [Pietermaritzburg: University of Natal Press, 1976], 157, Figure 84. Reproduced with permission from the University of KwaZulu-Natal Press.)

as their owners can afford to give them.[27] In this, they are unlike the true pariah dogs of India and the Middle East, which do usually live in semiwild packs outside human control.

Two kinds of indigenous African dogs have been adopted by Europeans and improved to the level of having Kennel Club status, the basenji from the equatorial forests of the Congo region and the Rhodesian ridgeback. The Rhodesian ridgeback, a large, tawny-colored dog with a ridge of hair running along the back in an opposite direction to the rest of the coat, is probably in part descended from indigenous Shona dogs crossed with European breeds such as the boarhound and Labrador retriever.

Figure 58. A pair of Africanis dogs in Zimbabwe. (Photo by author.)

Indigenous village dogs are everywhere in Africa today, and when not interbred with imported greyhounds and other breeds of European dogs, they are remarkably similar in conformation throughout their range. Toward the end of the twentieth century, a few dog owners and biologists realized that these dogs represent an ancient genetic lineage that should be preserved. In 1998 a meeting was held to discuss the conservation of traditional African dogs, and they were given the name Africanis, and the Africanis Society of Southern Africa was founded. Two books on the Africanis have followed, which are surely due to become classics in the new field of research into the history of ancient and traditional breeds of landraces.

Domesticates in the Americas

NORTH AMERICA WAS THE LAST HABITABLE CONTINENT TO BE REACHED BY ANATOMI-
cally modern humans, who probably walked there across the Bering Straits from Asia during
the Late Glacial Maximum (LGM), from around 13,000 to 10,000 years ago. At whatever
place they arrived in the inhospitable frozen tundra, these intrepid travelers must then have
moved south through ice-free corridors, but neither the archaeological nor the genetic evidence
has been able to establish the routes of their migration, and argument continues. However, it
is known from the work of Dennis O'Rourke and Jennifer Raff that the genetic diversity of
Native Americans is much reduced compared to that of their forbears in central and northeast
Asia, indicating that the first colonizers arrived in small founding groups.[1]

The time and means of arrival of humans in South America are even less well known than
in North America. It is always assumed that the migration occurred from north to south, and
yet in 1997 an occupation site was excavated at Monte Verde in Chile that was at first dated
to around 33,000 years ago, but is now known from radiocarbon dates to have been inhabited
around 14,000 years ago.[2] From where did these immigrants come?

THE ICE AGE FAUNA OF THE AMERICAS

During the long period of the last Ice Age and into the LGM, the Americas were inhabited
by faunal assemblages of large mammals and birds that today would be quite unrecognizable
to nearly everyone except the palaeontologists who research their skeletal remains. After the
melting of the ice, which began around 14,000 years ago, almost all these species became
extinct, and why this occurred on such a huge scale and in such a relatively short time has
been a subject of controversy among these palaeontologists for the past half a century. What
caused the sabre tooth cats, cheetahs, dire wolves, mammoths, mastodons, camels, giant
sloths, horses, vultures, and many more species within around thirty-five genera to die out?
As with the megafaunal extinctions in the rest of the world, two explanations have been considered possible: either the animals succumbed to the extreme global warming, which caused
dramatic environmental changes, or they were slaughtered by the human immigrants from
Asia with their large projectile, stone spears that characterize the Clovis industry. The full
causes for the American extinctions remain unexplained, but all the arguments and known
facts are discussed in a book edited by Gary Haynes, who (along with many others) has been
working on this subject for thirty years.[3] In simple terms, the causes may well have been the

environmental and climatic changes following the melting of the ice, combined with the hunting skills of the First Americans, as succinctly summarized by Steven Mithen in his global history, *After the Ice*.[4]

Whatever the causes, for the 9,000 or so years between the final extinction of the megafauna and the first European contact in the early second millennium CE, the mammalian fauna in the Americas was much restricted, although there were plenty of hoofed animals that could be hunted by humans. In North America, there were bison, several species of deer, wild sheep, and Rocky Mountain goats, with wolves, beavers, and rodents to provide furs, and many species of birds, including wild turkeys; in South America, there were camelids, peccaries, monkeys, and a great variety of carnivores, rodents, and birds. However, the First Americans did not all remain as hunter-gatherers; nations were established and were based on the same hierarchical systems as ancient and modern civilizations in the rest of the world. In Mesoamerica, cities were built, housing many thousands of inhabitants, and religious structures were erected that rivaled the pyramids of Egypt, but there was one great difference from civilizations everywhere else in the ancient world: although the Aztecs and Mayans cultivated plants, their economies were not based on the farming of livestock. The only domestic animals these nations had were the dog and the turkey.

DOMESTICATES IN MESOAMERICA AND NORTH AMERICA

Dogs

It is probable that the first hunter-gatherers to set foot on the continent of North America were accompanied by their dogs, who traveled with them from Asia. As the ice melted and the continents warmed, the founder populations of humans and dogs expanded rapidly, and their ways of life diversified as they adapted to the climate and environments of the lands through which they traveled. The humans developed nations with separate cultures, and their dogs evolved into breeds.

The earliest authentic description of the great variety of American aboriginal dogs was published by Glover Allen in 1920. Allen divided the kinds of North and South American dogs into seventeen groups: Eskimo, Plains Indian, Sioux, long-haired Pueblo, larger or common Indian, Klamath Indian, short-legged Indian, Klallam Indian, Inca, long-haired Inca, Patagonian, Mexican hairless, small Indian or Techichi, Hare Indian, Fuegian, short-nosed Indian, and Peruvian pug-nosed.[5] The earliest securely dated remains of dog in North America come from the Koster site in Illinois dated to about 8,500 years ago.[6]

It has been argued that the indigenous dogs of the Americas (those that were there before European contact) could have been locally domesticated from the American gray wolf. However, Jennifer Leonard and colleagues have shown that the genetic evidence indicates an Old World origin for all the specimens of Native American dogs that they analyzed. They sequenced the mtDNA from nineteen living Mexican hairless dogs, the Xoloitzcuintle, which is an ancient breed developed by the Mesoamericans long before European contact, and they also sequenced a large number of ancient bones from indigenous dog burials. They found that no ancient sequence matched those from American wolves.[7] Furthermore, as predicted by archaeozoologists for the last fifty years, the recent work of Bridgett von Holdt and colleagues indicates that all the past and present domestic dogs in the world, including the American, have ultimately descended from the gray wolf of the Middle East, although outbreeding with local wolves also occurred in the early history of specific lineages.[8] This could account for the previous findings of Ben Koop and Susan Crockford of genetic sequences from ancient dog remains that did match those of American wolves.[9]

Marion Schwartz has written a most comprehensive and original book, *The History of Dogs in the Early Americas*. She covered the arrival of dogs on the continents, the sites where their fossil remains have been found, the wide and varied ethnography of religious and secular rituals with dogs, their value as food, their use for haulage, and much else besides.

From archaeozoological reports, Schwartz summarizes the place of dogs in the Aztec, Olmec, and Mayan worlds, and it is clear that dogs were a common source of meat.[10] This has been evaluated in detail from the early levels of the Preclassic (1200 BCE–250 CE) site of Cuello in Belize, where the faunal remains showed that dogs had been bred for food and killed at the end of their first year of life, although white-tailed deer (*Odocoileus virginianus*) provided the most common source of meat, followed by turtles. The dogs were small, falling within the size range of the "small Indian dog or Techichi" of Allen's groups. The dogs did not belong to the hairless Xoloitzcuintle, since none of the mandibles showed the congenital lack of anterior teeth, which is genetically linked with hairlessness.[11]

Apart from the Xoloitzcuintle, which is known to have been kept purebred since ancient times, there are no known living breeds of indigenous Native American dogs, for all have been consciously and unconsciously interbred with immigrant European dogs, including the Eskimo dog and the modern breed known as the Native American Indian Dog. However, the many different breeds that were owned by the First Nations are well known from historical images and written accounts, as well as from their buried remains and from artifacts and clothing made from dog skins and hair.

Aboriginal peoples are not generally considered to have used artificial selection to develop breeds of domestic animals, and few examples are known from archaeology or history. However, a distinct breed known as the Salish wool dog has been described from the southwest coast of British Columbia by European accounts in the late 1700s. They were small, long-haired, white dogs that were bred exclusively for their thick, soft fur. The wool dogs were kept reproductively isolated from the larger village dogs, in packs, on offshore islands, with buried dried fish for food. They were owned by women who fed the dogs and sheared their thick fleeces several times a year. The wool was woven into Salish blankets, but with the arrival of sheep, brought to the country by Europeans, the wool dogs became extinct.[12]

As described by I. Lehr Brisbin Jr., there is a small population of semiwild dogs living in a remote area of South Carolina and called by him Carolina dogs. They are morphologically similar to dingoes, and Brisbin believes that these dogs have undergone no artificial selection and are direct descendents of the dogs that arrived with the first hunter-gatherers in the early Holocene.[13]

Turkeys

Wild turkeys only occur in North America, and there is only one species, *Meleagris gallopavo*, with six living subspecies, *M. g. gallopava*, *M. g. merriami*, *M. g. mexicana*, *M. g. silvestris*, *M. g. osceola*, and *M. g. intermedia*.

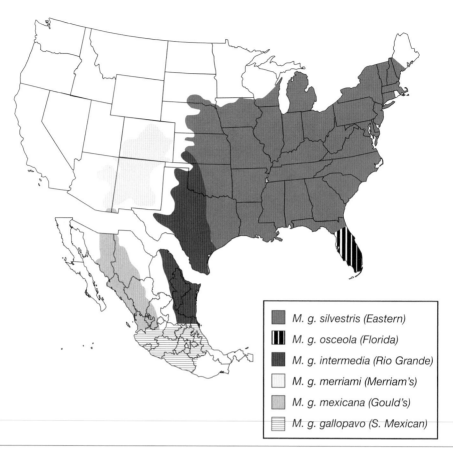

Figure 59. Modern distribution of the six living subspecies of wild turkeys, *Meleagris gallopavo*. (Source: Map from Camilla F. Speller et al., "Ancient Mitochondrial DNA Analysis Reveals Complexity of Indigenous North American Turkey Domestication," *Proceedings of the National Academy of Sciences* 107, no. 7 [16 February 2010], 2807–2812. Redrawn with permission from *Proceedings of the National Academy of Sciences*.)

All the world's domestic turkeys are descended from these wild birds, but although it has long been assumed that the Aztecs were the first to breed turkeys in captivity, until recently little has been known about the real phylogenetics of where and when this occurred.

With new genetic analysis, as with the history of so many other domestic species, the domestication of the turkey is revealed as a complex story, made more difficult by the past human transference of wild subspecies around the continent and the relic status of some subspecies. By the 1960s, it was known that there must have been two origins of domestication: in south-central Mexico, there are early remains of domestic turkeys dating from 200 BCE to 700 CE, and by the sixteenth century the birds were in Guatemala, Honduras, Nicaragua, Costa Rica, and some Caribbean islands. All these were originally identified (wrongly) as *M. g. osceola*. The second origin was in the southwestern United States (geographically very distant from Mexico), where the agricultural Pueblo people domesticated turkeys, and neither the 1960s data nor the new genetic evidence has provided any evidence that these birds played any part in the ancestry of modern domestic turkeys. The Pueblan turkeys from archaeological sites were also originally wrongly identified as the indigenous *M. g. merriami*.[14]

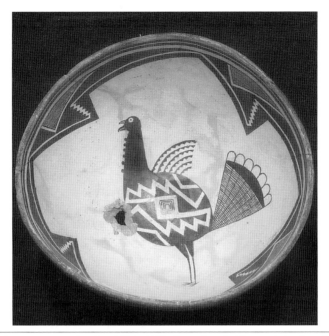

Figure 60. An Aztec turkey on a pottery bowl, c. 1000–1200. (Reproduced with permission from Eric Kaldahl/Amerind Foundation.)

The Pueblo people used turkey bones for tools and feathers for ornament and in weaving, and they used the spurs as arrowheads. Excavations have revealed that by 700–900 CE, large numbers of turkeys were kept in corrals, which have yielded vast quantities of turkey manure, eggshell fragments, and turkey bones.[15]

From new mtDNA analysis, Camilla Speller and colleagues have determined that two distinct groups of domestic turkeys are represented from the southwestern archaeological sites, and neither is the indigenous *M. g. merriami*. Instead, the phylogeographic, archaeological, and ancient DNA evidence indicates that from 200 BCE to 500 CE, turkeys were introduced from the east that genetically match *M. g. silvestris* and/or *M. g. intermedia*. Furthermore, there is a clear genetic distinction between these southwestern turkey remains and the center of domestication in the sites of south-central Mexico, where the Mexican turkey, *M. g. gallopava* (see figure 59), is now identified as the progenitor of turkey remains, and it is this subspecies that is closely related to all modern breeds of turkeys.[16]

Columbus may have been the first European to eat turkey meat when he landed on the coast of Honduras on 14 August 1502, and Cortés found large numbers of domestic turkeys in Mexico when he arrived there in 1519. Following the first Spanish contact, turkeys were taken to Europe in large numbers and spread extraordinarily rapidly, being well established on farms in Spain by 1530, and by the middle of the sixteenth century they had reached England, Denmark, and Norway, as described by R. D. Crawford.

The origin of the name "turkey" for these American birds is rather strange. The first Europeans to encounter them believed that the birds were a kind of Guinea fowl, which was also called the turkey fowl from its importation into Europe through Turkey, and this name stuck as the common name for the new bird. Linnaeus then gave *Meleagrus* as the Latin name for the turkey genus, which is also the genus name for the Guinea fowl.

According to Crawford, when Europeans began colonizing the east coast of North America, they brought turkeys back with them that had been bred in Europe, and turkeys were in Jamestown in Virginia in 1607.[17] James Harting in *The Birds of Shakespeare* gave several quotes from the plays as well as a rhyme about the introduction of the turkey to England, written in 1525: "Turkies, carps, hoppes, piccarell, and beere, / Came into England all in one yeare."[18]

A hundred years later, turkeys were commonplace and were regularly on the dinner menu, as in the diary of Samuel Pepys for 4 February 1659: "My wife killed her turkeys that came out of Zealand with my Lord, and could not get her maid Jane to kill anything at any time."

The most elegant image of a turkey to survive from the seventeenth century must be the Mogul painting from the period of Shah Jahan (1627–1658), which also demonstrates how quickly and how widely turkeys had spread round the world.

Figure 61. Mogul painting of a male turkey surrounded by flowers. (Photo © from the Fitzwilliam Museum, University of Cambridge.)

DOMESTICATES IN SOUTH AMERICA

Dogs

As in North America, it is probable that dogs accompanied the first human immigrants to the southern continent, and Marion Schwartz has summarized the fossil finds from the early sites. Dog remains are found together with those of camelids at Telarmachay and other rock shelters in the highlands of Peru, dating from 9000 to 2500 BCE.[19] From Fell's Cave in the southernmost tip of Chile there are ambiguous finds of canid remains that have been identified as domestic dog. However, the presumed very early date of 10,700–6,500 years ago has led to their identification being questioned, for there are several extinct and living wild species of indigenous canid, especially "foxes" belonging to the genus *Dusicyon*, that are possible candidates for these remains. But as flake tools and quantities of other animal remains have been excavated from the earliest levels at Fell's Cave, there is no reason why these early American hunters would not have had dogs.[20]

Camelids

In North America, the domestic dog, as well as fulfilling the usual roles of hunter, companion, guard, and scavenger, had to fulfill most of the functions, such as traction, that were carried out by domestic ungulates in the Old World. But in South America, the aboriginal people of the Andean region hunted and then domesticated the indigenous camelids, which enabled these people's descendents, the Incas, to live in the high mountains and helped to create the most powerful nation of the pre-Columbian Americas. The Inca Empire arose in the thirteenth century CE and collapsed in 1533 with the Spanish invasion and conquest.

There are two forms of domestic camelids living today, the llama (*Lama glama*) and the alpaca (*Vicugna pacos*), that are descended from the two living wild species, the guanico (*Lama guanicoe*) and the vicuna (*Vicugna vicugna*). The guanico is the largest wild hoofed mammal in South America today and ranges from the Andes to the plains of Patagonia. The vicuna is the smallest of the camelids and lives close to the snow line in the high Andes. It has a coat of the finest and lightest wool fiber in the world.

The faunal remains from a large number of archaeological sites in the Andes have been

Figure 62. Llamas near Cuzco, the historic capital of the Inca Empire. (Credit: Laverne Waddington, reproduced with permission.)

studied since the 1970s, notably by Jane Wheeler and colleagues, who were the first to suggest that a high mortality of newborn camelids in the assemblage from Telarmachay rock shelter in the Puna ecosystem of the Peruvian Andes indicated the management of domestic animals dating back to 6,000 years ago. Archaeozoological analysis of more than half a million animal bones indicated a shift from generalized hunting of guanaco, vicuna, and huemul deer (*Hippocamelus antisiensis*) 9,000 to 7,200 years ago to specialized hunting of guanaco and vicuna approximately 7,200 to 6,000 years ago. This was followed by control of early domestic camelids from 6,000 to 5,500 years ago, and finally to the establishment of a predominately herding economy beginning 5,500 years ago.[21] Wheeler's data, as well as the more recent reports of archaeozoologists from sites in the Andes, has been reviewed by Guillermo Mengoni Goñalons and Hugo Yacobaccio and highlights the complexity of camelid exploitation through the millennia. Despite the uncertainties, it seems probable that both llamas and alpacas had been domesticated by 5,000 years ago after several thousand years of exploitation as wild animals hunted for their meat and possibly also for their pelage, bones, and other tissues.[22]

Traditionally, the origins of the llama and the alpaca have been attributed to the wild guanaco, and it was claimed that the vicuna was never domesticated, but recent mitochondrial analysis has indicated that the alpaca is descended from the vicuna and not from the guanaco,[23] supporting Jane Wheeler's previous proposal based on changes in the morphology of the incisor teeth.[24] Accordingly, the Latin name of the alpaca should be changed from *Lama pacos* to *Vicugna pacos*.[25]

Living llamas and alpacas will fully hybridize and produce fertile offspring, which indicates

Figure 63. Ceramic alpaca, Moche culture,100 ce to 800 ce. (Reproduced with permission from the Larco Museum Collection.)

that their progenitors, the wild guanaco and vicuna, had a relatively recent evolutionary separation. Both the archaeozoological and the molecular evidence indicate that hybridization has a long history and followed the breakdown in the preconquest production of specialized breeds after European contact. During the first 100 years after the Spanish conquest, oral tradition claims that there was a loss of 80–90 percent of the herds as well as their herders. This would have led to the genetic bottleneck in the camelids in the sixteenth century that is evident in the poor quality of animals in the Andes today.[26]

Guinea Pigs

How the Guinea pig got its name has long been a puzzle. That these hystricomorph rodents were called "pigs" is not surprising, as with their fat little bodies and piglike squeals they resemble miniature suckling-pigs, but how the name "Guinea" became attached to them is more difficult to explain. One suggestion is that English-speaking travelers first saw these domesticated rodents on the northern coast of South America, which was associated with Guyana. The usual name in South America is "cuy" or "cui," and the accepted Latin name for the domestic form is *Cavia porcellus*, which was first ascribed to the Guinea pig in 1777 by Erxleben. Apart from the domestic Guinea pig, there are five wild species within the genus *Cavia* with the common name in English of cavies.[27]

The first European description of the Guinea pig was by the Spanish historian and writer Oviedo (1478–1557), who took part in the Spanish colonization of the Caribbean and wrote in 1547 that he saw animals called "cori" in Santo Domingo. As no cavy is indigenous to the Caribbean islands, he must have seen Guinea pigs that had been taken there by the Spanish invaders. By Oviedo's time, Guinea pigs had reached Europe, and in 1554 one was illustrated in Konrad Gesner's *Historiae animalium: Liber II de quadrupedibus oviparis*, where it is described as an "Indian rabbit."

Today, Guinea pigs are common domestic animals throughout the Andes in Peru, Columbia, Ecuador, and Bolivia, where they live as free-ranging commensals in the villages, providing meat much as they have done for thousands of years. Using molecular analysis, Angel Spotorno and colleagues found that *Cavia tschudii* from western South America is the wild species that is most closely related to the domestic form, *Cavia porcellus*, and they suggested a three-stage process for the domestication. The first ancient domestication of the pre-Columbian cavy was for the "criollo" (creole) breeds that are still found in the Andean countries. A second stage involved Europeans in the sixteenth century who took a few creole cavies to Europe and selectively bred them to produce the Guinea pigs that are today's worldwide pets and laboratory animals. The third stage has involved the modern selection of creole cavies for improved meat production.[28]

Muscovy Ducks

Wild Muscovy ducks (*Cairina moschata*) range from coastal Mexico through Central America to South America, reaching Peru in the west and Uruguay and northern Argentina in the southeast. They are tropical ducks that live in wetlands near forested areas, and they use trees for roosting and nesting. Although presumably the peoples of Central America knew of these ducks and probably hunted them, it was left to the South Americans to domesticate them. According to authors cited by G. A. Clayton, Muscovy ducks were probably first domesticated by Native Americans in Peru, where they were found in great numbers by the Spanish invaders. Marion Schwarz quotes a Spaniard as writing about indigenous people in the former Inca territory in 1548–1550: "Around their houses one sees many dogs, different from those of Spain, about the size of terriers, which they call *chonos*. They also raise many ducks."[29]

Like the turkey, Muscovy ducks reached Europe in the sixteenth century and very soon spread to North Africa. Like the Guinea pig, their name is a mystery since the duck has no original connection with Cairo, from which the genus name *Cairina* is derived, and has no connection with musk.[30]

THE IMPACT OF EUROPEAN INTRODUCTIONS OF DOMESTICATES AND THEIR FERAL DESCENDANTS IN THE AMERICAS

Dogs were the first domestic animals to be introduced in the New World with the First Americans, followed by an interval of around 9,000 years until the arrival of horses, donkeys, cattle, sheep, goats, pigs, chickens, and honey bees with the European invaders. In the words of Elinor Melville, the introduced ungulates "exploded into huge populations that in one way or another transformed the biological and social regimes of the New World."[31] When grazing animals inhabit land that has a surfeit of pasture and no predators, their numbers will increase exponentially until a population irruption results. This is followed by overgrazing, collapse of the carrying capacity of the land, and a population crash. The grazing pressure is removed, the pasture recovers, and the herbivores increase in numbers again, leading to a cycle of boom and bust also known as an irruptive oscillation. This can occur very rapidly over a few generations, and it partly explains the vast scale on which the introduced herbivores expanded over the New World grasslands.[32]

Melville describes the sheep irruption in Mexico in the 1570s as an example of the extraordinary rate of reproduction that was achieved by the introduced ungulates to the Americas and the devastating effect this had on the vegetation and thereby on the indigenous human population, causing what Alfred Crosby has defined, in a number of books, as ecological imperialism.[33] In the Valle del Mezquital the total number of sheep increased from an estimated 421,200 in the late 1550s to an estimated two million in 1565.[34]

After the extinction of the wild species of equids throughout the Americas, there was a gap of at least 9,000 years before the introduction of the first domestic horses by the Spanish invaders to Mexico (New Spain). Fear of the horse was the most effective weapon that the Spanish had against the Aztecs, and this fear played a large part in the total destruction of their empire. As Crosby recounts, once horses were established, although at first they were slow to breed, within a few years of 1550 there were 10,000 in the region of Querétaro, and they spread north with unimpeded progress.[35]

In South America, Pizarro traveled south to Peru on horseback, evoking terror and amazement in the Native Americans who had never seen animals like them, which "ran like the wind and killed people with their feet and mouths." Spanish settlers followed and traveled south, reaching the Argentine grasslands where the plains soon became "covered with escaped mares and horses in such numbers that they look like woods from a distance."[36] In the seventeenth century, when horses had increased to huge numbers, the way of life of the Native Americans of the grasslands changed from being based on horticulture to the eating of horsemeat and to hunting wild rheas, guanaco, and deer on horseback.[37]

The story of the introduction of horses to the Americas and their adoption by the indigenous peoples is very well known to all who have seen the plethora of films on "cowboys and Indians." It is also a story that exemplifies both an irruption and ecological imperialism—that is, the massive alteration and degradation in the ecological systems of great areas of the continents.

The vast herds of feral horses that were still living on the South American pampas in the early nineteenth century did not long survive the immigration of European landowners with their cattle and their firearms. However, on the Great Plains of North America, there are still thousands of feral horses, known as mustangs, as well as feral donkeys, called burros.[38]

The explosion of cattle populations in the Americas following the Spanish invasions equaled that of the horses and has been described by Crosby, who wrote, "At the end of the sixteenth century, the cattle herds in northern Mexico may have been doubling every fifteen years or so . . . and the great level plains, stretching endlessly and everywhere [were] covered with an infinite number of cattle." They were completely naturalized, as permanent a part of the fauna as the deer and the coyotes.[39]

The wild members of the pig family that are endemic to the New World are the peccaries, of which there are three species belonging to the family Tayassuidae: the white-lipped peccary (*Tayassu pecari*), the collared peccary (*T. tajacu*), and the chaco peccary (*Catagonus wagneri*). There were no pigs belonging to the family Suidae—that is, the wild boar and its descendant, the domestic pig—until they were introduced by humans. According to John Mayer and I. Lehr Brisbin Jr., who have written a detailed account of the feral swine throughout the Americas, the first pigs to reach the New World were taken to Hawaii by Polynesians during the first

Figure 64. Mustangs in the Great Basin, Nevada. (Source: Photo from the front jacket of Joel Berger, *Wild Horses of the Great Basin* [Chicago: University of Chicago Press, 1986]. Reproduced with permission from Joel Berger.)

century CE.[40] As with the other imported livestock, the Spanish were responsible for the first pigs to reach the mainland, and Crosby has written on their rapid expansion:

> Within a few years of Española's discovery, the number running wild was infinitos, and all the mountains swarmed with them. They spread to the other Greater Antilles and to the mainland in the 1490s. . . . They followed in the footsteps of Francisco Pizarro (who allegedly began life as a swineherd) and were soon doubling and redoubling their numbers in the area of the conquered Incan empire. . . . Every last one of these swarms of pigs, said the saintly Las Casas, were descendants of the eight pigs that Columbus had bought for seventy *maravedis* each in the Canaries and brought to Española in 1493.[41]

The only domesticated insect, the honey bee, *Apis mellifera*, whose earliest records are from ancient Egypt, arrived in Virginia in the early 1620s, and honey became a common food in North America from then onward, with the bees rapidly becoming naturalized and taken over by the Native Americans.[42]

Today, the numbers of domestic animals and their breeds in the Americas are perhaps as great or even greater than those in any other continent, and all but the turkeys, Muscovy ducks, Guinea pigs, camelids, and a few of the dogs are derived from Old World species.

Conclusions

FOR THE PAST 10,000 YEARS AND MORE, THE INTERACTION BETWEEN HUMANS AND animals has been evolving into an ever-closer relationship, which has moved away from that of predator and prey into a cooperative dependence from which neither can escape. Today, the majority of the world's populations of humans could not survive without the resources supplied by their domestic animals, and the innumerable breeds of domestic animals could not survive without the care and nourishment provided by their human owners. The gradual spread of this interdependence has followed the course of human history, although at different rates in different parts of the world, as described in the preceding chapters. The long story that has led to monopoly of the living world by the humans and domesticates of today is summarized here in three progressive phases that began at the end of the last Ice Age. The first phase is tied to settlement and the emergence of human societies; the second follows the spread of worldwide livestock husbandry and the founding of ancient empires; we live in the third phase, with the huge global expansion in numbers of both humans and domesticates that has led to radically altered methods of farming.

DOMESTICATES: THE EARLY PHASE

At least in part, it probably all began with people's affection for young animals, as suggested by Francis Galton, outlined in the Introduction, and backed up by the description of pet-keeping by hunter-gatherers written by Nicholas Guppy in his account of the Wai-Wai of the Amazon Basin:

> With baby creatures attention, tenderness, and care were undoubtedly the important factors. Creatures scarcely emerged from the womb or egg—parakeets the size of a finger-tip, baby humming birds no bigger than peas—were taken back to the villages and reared to adulthood: newly born mammals were suckled by the women, birds fed with pre-chewed cassava bread forced into their beaks, or even directly from the women's own mouths, in imitation of the mother birds, and if a creature were naturally shy or savage it was given to many people to handle, so that it became accustomed to human beings.[1]

As well as young animals, tamed in this early phase, wolves and many other wild animals that had a short flight distance,[2] and were therefore not afraid to come close to human habitations, would have been gradually enfolded into the human environment. These species included all the major domestic livestock of today, but not the highly nervous species such as gazelles and deer that rely on speed for escape from predators and have a long flight distance.

DOMESTICATES: THE PASTORAL PHASE

With settlement and the cultivation of plant crops, ownership of herd animals spread and became part of the way of life of people all over the world except on the continents of Australasia and North America. And it is indeed puzzling why kangaroos were not domesticated in Australia and why the bighorn sheep (*Ovis canadensis*) was not domesticated in North America.

Another puzzle, which has received little attention from archaeozoologists and archaeologists, is the possible rate of expansion of the numbers of herd animals after their introduction to new regions. For example, after sheep were first introduced to northern Europe, was there an irruption in their numbers as has been described by Elinor Melville for the introduction of sheep into Mexico in the sixteenth century and horses and cattle in both North and South America?[3] It could be envisaged that the early domesticated sheep, goats, cattle, pigs, and horses in Europe all bred successfully and produced huge, unmanageable populations in a remarkably short time. Yet there is little evidence for such irruptions in the archaeological record of the Neolithic. Perhaps the animals were too valuable, too well guarded, and their reproduction too well controlled, so there were few escapes and therefore no possibility for the establishment of feral populations. Or perhaps there were pockets of irruptions that have left no historical or living record apart from a few possible relics such as Britain's Exmoor ponies.

Following the establishment and spread of domestic livestock in the prehistoric periods, separate breeds evolved that were uniquely adapted to the biotope in which they lived; they were ecotypes that were as well adapted to their environment as the local wild fauna. The influence of these ancient breeds of domestic livestock is apparent in every part of the world, whether it be the Sahel, where herds of camels and goats range, or the landscape of Europe, which has been transformed over the past 5,000 years by ever-increasing numbers of grazing animals. From the Middle Ages, the species of trees allowed to grow in forests have been determined by the feeding of herds of pigs put out to pannage, while hillsides have been turned to pasture by the grazing of sheep and cattle, and moorlands have been created by overgrazing on poor soils.

Until the beginning of the Industrial Age, the majority of livestock herds were kept in relatively small numbers. Although they may often have been abused, overworked, or killed by cruel methods, the separate personalities of the animals were recognized; they were given names and treated as individuals. This phase, called here pastoral livestock husbandry, is personified in the many traditional English nursery rhymes, some of which date from the Middle Ages, for example:

> Little boy blue,
> Come blow on your horn,
> The sheep's in the meadow,
> The cow's in the corn;
> But where is the little boy
> Who looks after the sheep?
> He's under a haycock,
> Fast asleep.

Over the centuries, livestock breeds were moved around countries and even over continents, which temporarily destroyed their environmental adaptation. For example, the Romans are believed to have brought white sheep and possibly their own breeds of cattle to Britain, but these animals in turn became adapted to the new local conditions of soil and climate. Local breeds (also known as landraces) survive in most countries today, although in greatly reduced numbers, and realization of their historical and genetic value has led to the founding of many societies for their conservation. Examples are the Rare Breeds Survival Trust in the United Kingdom, the Equus Survival Trust in North America, and the Farm Animal Conservation Trust in South Africa.

For the present, in many parts of the world away from the densely populated industrial centers, pastoral farming and livestock herding are still the basis of the economy and the mainstay of rural communities. From the reindeer herders of the Arctic to the Maasai cattle herders of East Africa, in marginal zones this ancient way of life is likely to continue for many generations to come. However, the "old way"—as related, for example, so empathetically by the anthropologist Elizabeth Marshall Thomas—is unlikely to occur anywhere today.[4] She describes the relationship of the Kalahari herders (whom she knew in her youth) with their cattle and the local lions: the inheritance of perhaps thousands of years of learned behavior, or culture, that enabled people, livestock, and predators to live in harmony.

DOMESTICATES IN THE MODERN INDUSTRIAL WORLD

The change from the pastoral herding of livestock to modern farming in the Western world began with the impetus for "improvement" in Europe in the eighteenth century when the Industrial Revolution made it necessary to increase the quantity and quality of meat, tallow, wool, and leather for the rapidly expanding urban populations. "Improvement" means increasing the productivity of a breed of domestic animals by artificial selection, but the early "improvers" knew nothing of genetics or evolution. They did not realize that by crossbreeding animals from different localities, they were destroying breeds that had taken hundreds of years to evolve adaptations to particular environmental conditions. Today, the legacy of the "improvers" can still be seen in the many schemes that have attempted to improve cattle all over the world. In Africa, for example, where the native breeds have survived the onslaughts of dangerous local parasites such as trypanosomes, as well as severe periodic droughts for the past 2,000 years, European breeds such as the Friesian and the Hereford have been imported and crossed with local breeds such as the Boran in Ethiopia. The humped Boran may not have a large milk yield, but these cattle are perfectly adapted to the semidesert region of the Sahel in that they only need to drink once every three days. They cannot be improved by introgression from European cattle that have evolved in a totally different environment and climate. In the short term this "improvement" will lead to increased productivity, but there is a loss of the unique genetic constitution of the local breed, and susceptibility to stress and to disease means that greater protection and immunization are needed for the new, valuable but vulnerable herds. Such schemes for supposed improvement often appear to be quite irrational, except that they were carried out by European colonists who believed that what is familiar must be the best.

The arguments against "improvement" for its destruction of adaptation to local environments are, sadly, today only of historical interest, for what rules now is how to feed the global markets of the industrial world as economically and efficiently as possible. This inevitably leads away from pastoral farming along the easier path to mechanized livestock production, which means that the animals, whether they are pigs, cattle, or chickens, are no longer seen as individual sentient beings but as rows of food-producing elements to be cultivated and then harvested in enormous numbers.

Cruelty is difficult, if not impossible, to avoid with the keeping of large numbers of animals in small indoor spaces—that is, in factory farms, which have spread rapidly worldwide since the 1950s. There have been many impassioned writings against the intensive indoor methods of keeping animals for food—for example, by Compassion in World Farming (a society founded in 1967),[5] and as described by Jonathan Safran Foer in his book, *Eating Animals.*[6] He asks how can anyone agree to keep 33,000 chickens in a shed 45 feet wide by 490 feet long without total degradation to the owner and the birds?

The sad situation is that whereas it is relatively easy to raise funds to protect endangered wild animals such as elephants or tigers from being killed by poachers, it is quite another matter to raise funds for improving the welfare and manner of slaughter of domestic animals in factory farms. The majority of people, worldwide, just do not want to know how the meat and eggs that they eat and the milk that they drink get into the packages in their supermarket. And as the human population expands, so will the numbers of livestock have to increase with all the associated horrors in their short lives and cruel deaths.

The future looks bleak, but ways must be found to improve the new methods of farming, which are here to stay. One method is already under way, this being the production of breeds that are adapted as physically and mentally as possible to life in the factory farm. Just as in the Middle Ages, when pigs had slender bodies, long legs, and a relatively long flight distance in adaptaton to life under the system of pannage in forests, so now breeds of pigs that live their lives in small indoor stalls must be heavy-bodied and have short legs and passive temperaments with no flight distance.

But no amount of breeding for life in a factory farm can alleviate the dramatic changes that this new system of food production brings. So far, factory farming in the small islands of the United Kingdom has been more or less restricted to poultry, but 2010 has seen proposals for huge industrial complexes that could house thousands of animals, including a project for a major new form of milk production at Nocton, near Lincoln in the east of England. In a forty million pound scheme, a superdairy would have eight sheds with cubicles housing 8,000 cows that would lie on sand and be automatically milked to produce 430,000 pints of milk every day. The factory would have an anaerobic digester to produce gas from the slurry, which would generate enough electricity for running the factory as well as more than 2,000 homes. Concerted objections on many counts, including animal welfare, have forced the temporary withdrawal of this scheme, which to many people must seem like science fiction. But superfarms are becoming increasingly common in America and elsewhere, and it is hard not to see this as the way of farming for the future everywhere. However, the choices facing agriculture and farming practice in the modern world are not only about successful production and sustainability, they are also moral and ethical. The universal aim should be respect for the land and all its inhabitants, realization that every domesticated animal is a sentient being with a personality of his or her own, and remembrance of Charles Kingsley's "Mrs. do-as-you-would-be-done-by."[7]

DOMESTICATES IN THE FUTURE: SPECULATION

If only appropriate welfare of factory-farmed animals could be given the priority that humanity demands, this new form of food production could have advantages for natural biotopes and the biodiversity of wildlife. The ending of the age-old system of pastoral husbandry might release large areas of land from grazing, much of which would be unsuitable for arable crops. This land could provide a new resource in herds of free-ranging, feral animals as it does with the mustangs and burros of the Great Plains of North America, and in the arid lands of Australia where flocks of feral goats are rounded up for the export of meat.

Nomenclature of the Domestic Animals and Their Wild Progenitors

THE TENTH EDITION OF THE *SYSTEMA NATURAE*, WRITTEN BY LINNAEUS AND PUB-lished in 1758, is internationally accepted as the basis for formal zoological nomenclature. Written in Latin, Linnaeus gave names, descriptions, and localities to more than 4,000 organisms, among which were included all the common domestic animals.[1] When Linnaeus was familiar with both the wild and the domestic form of a species and they looked alike, as with his native reindeer, he gave them the same name, *Cervus tarandus* (now called *Rangifer tarandus*), for example. On the other hand, because he failed to see the relationship between the wolf and the dog, he gave them separate species names, *Canis lupus* and *Canis familiaris*, respectively. With yet others—for example, goats and sheep—he had no knowledge of the wild ancestor, and so he named only the domestic form. Implicit in these three examples is the still unresolved problem for many archaeozoologists of whether domestic animals should be treated as taxonomically identical with their wild progenitors.

Domestic animals, although they are fully fertile with their wild progenitors, do not normally breed with them. They usually live and breed as separate, reproductively isolated populations, and therefore, under the definition of the biological species concept, they should be considered as distinct species (see the Introduction), which for purposes of clarity require species names that are consistent and which, by tradition and common usage, are known to everyone. The obvious names to use are those that hold priority by being the oldest, but difficulties of identification arise when these names are also used for the wild progenitor. This is because there is a widely accepted premise that names based on descriptions of domestic mammals should not be used for wild species; for example, *Ovis aries*, which was named by Linnaeus on a domestic sheep, should not be the name used also for the wild species. This principle has received added support within recent decades from research into the molecular history of a number of domestic species, including sheep, which indicates that they are the descendants of more than one wild, genetically distinct subspecies. Therefore, an application was submitted to the International Commission for Zoological Nomenclature to authorize use of the next available name for the wild species, and this was approved in 2003.[2] Thus the official name for the wild ancestor of the domestic sheep is now *Ovis orientalis*. Further details on the nomenclature of domestic species can be found in the 2004 article by Gentry, Clutton-Brock, and Groves.[3]

LATIN NAMES OF THE DOMESTIC ANIMALS AND THEIR
WILD PROGENITORS AS USED IN THIS BOOK

Not all domestic species are listed here, and the commensal species—for example, rats and mice—are not included. The distributions are broad assessments of what was probable for the wild species at the time of their first domestication.

Wild Progenitor	*Domestic species*
Mammals	
ORDER LAGOMORPHA	
Oryctolagus cuniculus (L), wild rabbit	*Oryctolagus cuniculus* (L), domestic rabbit
Spain and possibly south of France and parts of northwest Africa	
ORDER RODENTIA	
Cavia tschudii Fitzinger, 1857, cavy	*Cavia porcellus* (L), Guinea pig
South America	
ORDER CARNIVORA	
Canis lupus L, wolf	*Canis familiaris* L, dog
Widespread throughout the Northern Hemisphere except Africa	
Mustela putorus L, European polecat	*Mustela furo* L, domestic ferret
Europe	
Felis silvestris Schreber, 1777, wild cat	*Felis catus* L, domestic cat
Europe, Asia, Africa	
ORDER PERISSODACTYLA (odd-toed ungulates)	
Equus ferus Boddaert, 1785, wild horse	*Equus caballus* L, domestic horse
Southern Russia and Central Asia	
Equus africanus (Fitzinger, 1857) wild ass	*Equus asinus* L, domestic donkey
North Africa and possibly Western Asia	
ORDER ARTIODACTYLA (even-toed ungulates)	
Sus scrofa L, wild boar	*Sus domesticus* Erxleben, 1777, domestic pig
Europe, Asia, and North Africa	
Lama guanicoe (Muller, 1776), guanaco	*Lama glama* (L), llama
South America, semideserts and high altitudes	
Vicugna vicugna, vicugna	*Vicugna pacos* (L), alpaca
Andes, South America	
Camelus—species not known	*Camelus dromedarius* L, dromedary or one-humped camel
Desert, western Asia, or possibly North Africa	
Camelus ferus Przewalski, 1883, wild Bactrian camel	*Camelus bactrianus* L, domestic Bactrian or two-humped camel
Central Russia and Asia, dry steppe and desert	

Rangifer tarandus (L) reindeer, caribou	*Rangifer tarandus* (L), domestic reindeer
Northern Europe and Asia, and North America	
Bubalus bubalis (L), river and swamp buffaloes	*Bubalus arnee* (Kerr, 1792), water buffalo
India, southern Asia, and possibly western Asia, wet areas	
Bos primigenius Bojanus, 1827, aurochs or giant ox (extinct)	*Bos taurus* L, European domestic cattle
Europe, Asia, and North Africa	
Bos namadicus Falconer, 1853, Indian "aurochs" (extinct)	*Bos indicus* L, Indian humped cattle or zebu
Bos mutus (Przewalski 1883) yak	*Bos grunniens* L, domestic yak
Mountains of Tibet, Nepal, and the Himalayas	
Bos gaurus H. Smith, 1827, gaur	*Bos frontalis* Lambert,1804, mithan or gayal
Bos javanicus d'Alton, 1823, banteng	*Bos javanicus* d'Alton, 1823, Bali cattle
Borneo and islands of Southeast Asia	
Capra aegagrus Erxleben, 1777, bezoar goat	*Capra hircus* L, domestic goat
Mountains of western Asia	
Ovis orientalis Gmelin, 1774, Asiatic and European mouflon	*Ovis aries* L, domestic sheep
Mountains of western Asia	

Birds

ORDER COLUMBIFORMES

Columba livia Gmelin, 1789, rock dove	*Columbia livia*, domestic pigeon

ORDER GALLIFORMES

Gallus gallus (L), red jungle fowl	*Gallus domesticus*, domestic chicken
Southeast Asia	
Gallus sonneratii Temminck, 1813, grey jungle fowl	
India	
Meleagris galloparvo L, wild turkey	*Meleagris galloparvo* L, domestic turkey
Mexico and Central America	
Numida meleagris (L), helmeted Guinea fowl	*Numida meleagris* (L), domestic Guinea fowl
Africa	
Pavo cristatus L, wild peafowl	*Pavo cristatus* L, domestic peacock
India	

ORDER ANSERIFORMES

Cairina moschata (L), Muscovy duck	*Cairina moschata* (L), Muscovy duck
Central and South America	

Fish

ORDER CYPRINIFORMES

Carassius auratus (L), gibel carp	*Carassius auratus* (L), goldfish
Eastern Europe to Siberia	

Insect

ORDER HYMENOPTERA

Apis mellifera L, wild honeybee   *Apis mellifera* L, domestic honeybee

Europe and Africa

Notes

FOREWORD

1. M. Appleby, *Eating Our Future: The Environmental Impact of Industrial Animal Agriculture* (London: World Society for the Protection of Animals, 2008).

INTRODUCTION

1. Aristotle, *Aristotle: Generation of Animals*, Loeb Classical Library, trans. A. L. Peck (Cambridge, Mass.: Harvard University Press, 1990), p. xli.
2. As a legacy from the Greek philosophers and from medieval times, it was believed that a continuous chain of creation extended from the inanimate world of nonliving matter such as earth and stones, through the animate world of plants, zoophytes, and the lowest forms of animal life, upward to the quadrupeds and eventually through humans to the realms of angels and finally to the Christian God. This belief also entailed the view that just as nothing new could be created, neither could anything be exterminated, since this would counteract the will of God. See Juliet Clutton-Brock, "Aristotle, the Scale of Nature, and Modern Attitudes to Animals," in *Humans and Other Animals*, ed. Arien Mack (Columbus: Ohio State University Press, 1995), 5–24.
3. Jared Diamond, *The Third Chimpanzee: The Evolution and Future of the Human Animal* (New York: Harper Perennial, 2006).
4. Thomas Henry Huxley, *Man's Place in Nature and Other Essays* (London: Dent and Dutton, Everyman, 1911).
5. Charles Darwin, *On the Origin of Species by Means of Natural Selection or the Preservation of Favoured Races in the Struggle for Life* (London: John Murray, 1859); Charles Darwin, *The Variation of Animals and Plants under Domestication*, 2nd ed. (London: John Murray, 1890).
6. Sherwood L. Washburn and C. S. Lancaster, "The Evolution of Hunting," in *Man the Hunter*, ed. by Richard B. Lee and Irven Devore (Chicago: Aldine, 1968), 293–303.
7. Sarah Blaffer Hrdy, *Mothers and Others: The Evolutionary Origins of Mutual Understanding* (Cambridge, Mass.: Harvard University Press, 2009). Hrdy uses the term "cooperative breeding," but in zoology this term is restricted to animals such as termites and naked mole rats that live socially with one breeding female and a large number of "workers." It is therefore preferable to use the term "communal breeding" for animals (for example, meerkats) and humans that share the care of their young. T. H. Clutton-Brock, personal communication with author, 30 May 2009.
8. Francis Galton, "Domestication of Animals," in *Enquiries into Human Faculty and Its Development*, 2nd ed. (London: Dent and Dutton, Everyman, 1907), 173–193.
9. Galton, "Domestication of Animals," 181.
10. 2 Samuel 12:3.

11. Galton, "Domestication of Animals," 187.

12. X.-B. Jin and A. L. Yen, "Conservation and the Cricket Culture in China," *Journal of Insect Conservation* 2 (1998): 211–216.

13. Ann Thwaite, *Edmund Gosse: A Literary Landscape* (Stroud, Gloucestershire: Tempus, 2007), 209.

14. Galton, "Domestication of Animals," 194.

15. Darwin, *Variation of Animals and Plants under Domestication*, 2:177–178.

16. Juliet Clutton-Brock, *A Natural History of Domesticated Mammals*, 2nd ed. (Cambridge: Cambridge University Press/Natural History Museum, 1999) 32; Juliet Clutton-Brock, "How Domestic Animals Have Shaped the Development of Human Societies," in *A Cultural History of Animals in Antiquity*, ed. by Linda Kalof (New York: Berg, 2007), 71.

17. Clutton-Brock, *Natural History of Domesticated Mammals*, 30.

18. Ernst Mayr, *Animal Species and Evolution* (Cambridge, Mass.: Belknap 1966), 19.

19. Achilles Gautier, "What's in a Name?" in *Skeletons in Her Cupboard: Festschrift for Juliet Clutton-Brock*, ed. by Anneke Clason, Sebastian Payne, and Hans-Peter Uerpmann (Oxford: Oxbow Monograph 34, 1993), 91–98.

20. International Commission on Zoological Nomenclature, "Usage of 17 Specific Names Based on Wild Species Which Are Predated by or Contemporary with Those Based on Domestic Animals (Lepidoptera, Osteichthyes, Mammalia): Conserved," *Bulletin of Zoological Nomenclature* 60, no. 1 (2003): 81–84; Anthea Gentry, Juliet Clutton-Brock, and Colin Groves, "The Naming of Wild Animal Species and Their Domestic Derivatives," *Journal of Archaeological Science* 31 (2004): 645–651.

21. Joel Berger, *Wild Horses of the Great Basin* (Chicago: University of Chicago Press, 1986).

22. Kenneth Page Oakley, *Man the Toolmaker* (Chicago: University of Chicago Press, 1968).

23. Frans de Waal, *The Ape and the Sushi Master: Cultural Reflections by a Primatologist* (New York: Penguin Books, 2001).

24. de Waal, *The Ape and the Sushi Master,* 31.

25. Tim Ingold, "From Trust to Domination: An Alternative History of Human-Animal Relations," in *Animals and Human Society: Changing Perspectives*, ed. by Aubrey Manning and James Serpell (New York: Routledge, 1994) 1–22; Juliet Clutton-Brock, "The Unnatural World: Behavioural Aspects of Humans and Animals in the Process of Domestication," in Manning and Serpell, *Animals and Human Society*, 23–35.

26. Sadahiko Nakajima, Mariko Yamamoto, and Natsumi Yoshimoto, "Dogs Look Like Their Owners: Replications with Racially Homogenous Owner Portraits," *Anthrozoös* 22, no. 2 (2009): 173–181.

27. Jared Diamond, *Guns, Germs and Steel: A Short History of Everybody for the Last 13,000 Years* (London: Vintage, 1998), chapter 9; Juliet Clutton-Brock, "The Domestication Process: The Wild and Tame," in *Encyclopedia of Human-Animal Relationships*, ed. by Marc Bekoff (Westport, Conn.: Greenwood Press, 2007), 639–643.

28. Stephen Budiansky, *The Covenant of the Wild: Why Animals Chose Domestication* (New York: William Morrow, 1992).

29. Diamond, *Guns, Germs and Steel*, chapter 9.

30. The flight distance of a species is the average distance that individual animals within that species will allow between themselves and a potential predator before they flee. Species with a short flight distance allow potential predators or humans to approach closer than species that have a long flight distance.

31. Clutton-Brock, *Natural History of Domesticated Mammals*, 32–38; Helmut Hemmer,

Domestication: The Decline of Environmental Appreciation (Cambridge: Cambridge University Press, 1990).

32. Philip L. Armitage and Juliet Clutton-Brock, "A System for Classification and Description of the Horn Cores of Cattle from Archaeological Sites," *Journal of Archaeological Science* 3 (1976): 329–348.

33. D. Belyaev and L. N. Trut, "Some Genetic and Endocrine Effects of Selection for Domestication in Silver Foxes," in *The Wild Canids*, ed. by M. W. Fox (New York: Van Nostrand Reinhold, 1975), 416–426; L. N. Trut, "Early Canid Domestication: The Farm-Fox Experiment," *American Science* 87 (1999): 160–169.

34. Keith Dobney and G. Larson, "Genetics and Animal Domestication: New Windows on an Elusive Process," *Journal of Zoology* 269 (2006): 261–271; Susan J. Crockford, *Rhythms of Life: Thyroid Hormone and the Origin of Species* (Bloomington, Ind.: Trafford, 2003).

CHAPTER 1. EURASIA AFTER THE ICE

1. Rudolf Musil, "Domestication of Wolves in Central European Magdalenian Sites," in *Dogs through Time: An Archaeological Perspective*, ed. by Susan Janet Crockford (Oxford: BAR International Series 889, 2000), 21–28; M. V. Sablin and G. A. Khlopachev, "The Earliest Ice Age Dogs: Evidence from Eliseevichi," *Current Anthropology* 43 (2002): 795–799.

2. C. Vilà et al., "Multiple and Ancient Origins of the Dog," *Science* 276 (13 June 1997): 1687–1689.

3. Mietje Germonpréa et al., "Fossil Dogs and Wolves from Palaeolithic Sites in Belgium, the Ukraine and Russia: Osteometry, Ancient DNA and Stable Isotopes," *Journal of Archaeological Science* 36, no. 2 (2009): 473–490.

4. Germonpréa et al., "Fossil Dogs and Wolves from Palaeolithic Sites," 474.

5. Robert K. Wayne, Jennifer A. Leonard, and Carles Vilà, "Genetic Analysis of Dog Domestication," in *Documenting Domestication: New Genetic and Archaeological Paradigms*, ed. by Melinda A. Zeder, Daniel G. Bradley, Eve Emshwiller, and Bruce D. Smith (Berkeley: University of California Press, 2006), 279–293.

6. Antony J. Sutcliffe, *On the Track of Ice Age Mammals* (London: British Museum, 1985).

7. Steven Mithen, *After the Ice: A Global Human History, 20,000–5000 b.c.* (London: Weidenfeld and Nicolsen, 2003), 8.

8. Elaine Turner, "Results of a Recent Analysis of Horse Remains Dating to the Magdalenian Period at Solutré, France," in *Equids in Time and Space: Papers in Honour of Vera Eisenmann*, ed. by Marjan Mashkour (Oxford: Oxbow Books, 2006), 70–89.

9. Antonella Cinzia Marra, "Pleistocene Mammals of Mediterranean Islands," *Quaternary International* 129 (2005): 5–14.

10. J. S. Kopper and W. H. Waldren, "Balearic Prehistory: A New Perspective," *Archaeology* 20 (1967): 108–115.

11. Richard Burleigh and Juliet Clutton-Brock, "The Survival of *Myotragus balearicus* Bate, 1909, into the Neolithic on Mallorca," *Journal of Archaeological Science* 7 (1980): 385–388.

12. S. J. M. Davis, *The Archaeology of Animals* (London: Batsford, 1987), 118–125.

13. Homer, *Odyssey*, trans. by E. V. Rieu (England: Penguin Books, 1946), 139–154.

14. Adrienne Mayor, *The First Fossil Hunters: Paleontology in Greek and Roman Times* (Princeton, N.J.: Princeton University Press, 2000), 5–6.

15. Mithen, *After the Ice*, 108–109 (map).

16. Paul A. Mellars, "The Palaeolithic and Mesolithic," in *British Prehistory: A New Outline*, ed. by Colin Renfrew (London: Duckworth, 1974), 77–81; Jos Deeben and Nico Arts, "From Tundra Hunting to Forest Hunting: Later Upper Palaeolithic and Early Mesolithic," in *The Prehistory of the Netherlands*, vol. 1, ed. by L. P. Louwe Kooijmans et al. (Amsterdam: Amsterdam University Press, 2005), 151–153.

17. F. Poplin, "Origine du mouflon de Corse dans une nouvelle perspective paléontologique: par marronage," *Annales Génétique et de Sélection Animale* 11, no. 2 (1979): 133–143.

18. Juliet Clutton-Brock, "Origins of the Dog: Domestication and Early History," in *The Domestic Dog: Its Evolution, Behaviour, and Interactions with People*, ed. by James Serpell (Cambridge: Cambridge University Press, 1995), 13.

19. Martin Street, "Ein frühmesolithischer Hund und Hundeverbiß an Knochen vom Fundplatz Bedburg-Königshoven, Niederrhein," *Archäologische Informationen* 12 (1989): 203–215.

20. J. G. D. Clark, *Excavations at Star Carr* (Cambridge: Cambridge University Press, 1954), 176–177, plates XXII–XXIV.

21. Clutton-Brock, "Origins of the Dog," 13; Rick J. Schulting and Michael P. Richards, "Dogs, Divers, Deer and Diet: Stable Isotope Results from Star Carr and a Response to Dark," *Journal of Archaeological Science* 36 (2009): 498–503.

22. Juliet Clutton-Brock, and Noe-Nygaard, "New Osteological and C-isotope Evidence on Mesolithic Dogs: Companions to Hunters and Fishers at Star Carr, Seamer Carr and Kongemose," *Journal of Archaeological Science* 17 (1990): 643–653.

23. Schulting and Richards, "Dogs," 499–500.

24. Clark, *Excavations at Star Carr,* 70–95.

25. P. C. Woodman, *The Mesolithic in Ireland* (Oxford: BAR British Series 58, 1978), 135–136.

26. Piers Vitebsky, *Reindeer People* (London: Harper Perennial, 2005), 23.

27. Tim Ingold, *Hunters, Pastoralists and Ranchers: Reindeer Economies and Their Transformations* (Cambridge: Cambridge University Press, 1980), 71.

28. Ingold, *Hunters, Pastoralists and Ranchers,* 58–65.

29. P. S. Zhigunov, ed., *Reindeer Husbandry* (Jerusalem: Israel Program for Scientific Translations, 1968), 61.

30. Vitebsky, *Reindeer People*, 28–33.

31. J.-D. Vigne, J. Guilaine, K. Debue, L. Haye, and P. Gérard, "Early Taming of the Cat in Cyprus," *Science* (Brevia) 304, no. 5668 (2004): 259.

32. S. J. M. Davis, *The Archaeology of Animals* (London: Batsford, 1987), 133–134.

33. L. Chaix, A. Bridault, and R. Picavet, "A Tamed Brown Bear (*Ursus arctos* L.) of the Late Mesolithic from La Grande-Rivoire (Isère, France)?" *Journal of Archaeological Science* 24, no. 12 (1997): 1067–1074; Juliet Clutton-Brock, *A Natural History of Domesticated Mammals*, 2nd ed. (Cambridge: Cambridge University Press, 1999), 8.

34. J. E. King, "Mammal Bones from Khirokitia and Erimi," in *Khirokitia: Final Report on the Excavation of a Neolithic Settlement in Cyprus on Behalf of the Department of Antiquities, 1936–1946*, ed. by P. Dikaios (Oxford: Oxford University Press, 1953), 431–437.

CHAPTER 2. SETTLEMENT AND DOMESTICATION IN EURASIA

1. There are three species of gazelles in western Asia, the dorcas gazelle, *Gazella dorcas*; the Arabian gazelle, *Gazella gazella*; and the goitered gazelle, *Gazella subgutturosa*.
2. James Mellaart, *The Neolithic of the Near East* (London: Thames and Hudson, 1975), 27.
3. Ofer Bar-Yosef and François Valla, eds., *The Natufian Culture in the Levant* (Ann Arbor, Mich.: International Monographs in Prehistory Archaeological Series 1, 1991).
4. S. J. M. Davis and F. R. Valla, "Evidence for Domestication of the Dog 12,000 Years Ago in the Natufian of Israel," *Nature* 276, no. 5688 (1978): 608–610.
5. Juliet Clutton-Brock, "Origins of the Dog: Domestication and Early History," in *The Domestic Dog: Its Evolution, Behaviour, and Interactions with People*, ed. by James Serpell (Cambridge: Cambridge University Press, 1995), 10.
6. E. F. F. Valla, "Two New Dogs, and Other Natufian Dogs, from the Southern Levant," *Journal of Archaeological Science* 24, no. 1 (1997): 65–95.
7. Tamar Dayan, "Early Domesticated Dogs of the Near East," *Journal of Archaeological Science* 21 (1994): 633–640.
8. Carlos A. Driscoll et al., "Near Eastern Origin of Cat Domestication," *Science* 317 (2007): 519–523; Carlos A. Driscoll, Juliet Clutton-Brock, Andrew C. Kitchener, and Stephen J. O'Brien, "The Taming of the Cat," *Scientific American* 300, no. 6 (June 2009): 56–63.
9. Thomas Cucchi and Jean-Denis Vigne, "Origin and Diffusion of the House Mouse in the Mediterranean," *Human Evolution* 21 (2006): 95–106.
10. J.-D. Vigne, J. Guilaine, K. Debue, L. Haye, and P. Gérard, "Early Taming of the Cat in Cyprus," *Science* 304, no. 5668 (2004): 259.
11. Juliet Clutton-Brock, *The British Museum Book of Cats Ancient and Modern* (London: British Museum Press and Natural History Museum, 1988).
12. David R. Harris, ed., *The Origins and Spread of Agriculture and Pastoralism in Eurasia* (London: UCL Press, 1996); Melinda A. Zeder, et al., eds., *Documenting Domestication: New Genetic and Archaeological Paradigms* (Berkeley: California University Press, 2006).
13. K. M. Kenyon, *Digging Up Jericho* (London: Ernest Benn, 1957); Juliet Clutton-Brock, "The Mammalian Remains from the Jericho Tell," *Proceedings of the Prehistoric Society* 45 (1979): 135–157.
14. Andrew Garrard, Susan Colledge, and Louise Martin, "The Emergence of Crop Cultivation and Caprine Herding in the 'Marginal Zone' of the Southern Levant," in Harris, *Origins and Spread of Agriculture*, 208–221.
15. Clutton-Brock, "Mammalian Remains from the Jericho Tell," table 2, 138.
16. Hans-Peter Uerpmann, "Animal Domestication: Accident or Intention?" in Harris, *Origins and Spread of Agriculture*, 235.
17. G. Luikart, H. Fernandez, M. Mashkour, P. R. England, and P. Taberlet, "Origins and Diffusion of Domestic Goats Inferred from DNA Markers Example Analyses of mtDNA, Y Chromosome, and Microsatellites," in Zeder et al., *Documenting Domestication*, 294–305.
18. Juliet Clutton-Brock, *A Natural History of Domesticated Mammals*, 2nd ed. (Cambridge: Cambridge University Press, 1999), 70–76.
19. Clutton-Brock, *A Natural History of Domesticated Mammals*, 72–73.
20. Yutaka Tani, "The Geographical Distribution and Function of Sheep Flock Leaders: A Cultural Aspect of the Man-Domesticated Animal Relationship in Southwestern Eurasia," in *The Walking*

Larder: Patterns of Domestication, Pastoralism and Predation, ed. by Juliet Clutton-Brock (London: Unwin Hyman: 1990), 185–199.

21. Clutton-Brock, "Mammalian Remains from the Jericho Tell," table 2, 138.

22. Gregor Larson, Umberto Albarella, Keith Dobney, and Peter Rowley-Conwy, "Current Views on *Sus* Phylogeography and Pig Domestication as Seen through Modern mtDNA Studies," in *Pigs and Humans: 10,000 Years of Interaction*, ed. by Umberto Albarella, Keith Dobney, Anton Ervynck, and Peter Rowley-Conwy (Oxford: Oxford University Press, 2007), 30–31.

23. The last recorded live aurochs (*Bos primigenius*), a female, was killed in 1627 in Poland, in the Jaktorów Forest.

24. Clutton-Brock, "Mammalian Remains from the Jericho Tell," table 2, 138.

25. James Mellaart, *Çatal Hüyük: A Neolithic Town in Anatolia* (London: Thames and Hudson, 1967).

26. Ian Hodder, *Çatalhöyük: The Leopard's Tale, Revealing the Mysteries of Turkey's Ancient "Town"* (London: Thames and Hudson, 2006).

27. Mellaart, *Neolithic of the Near East*, 98–111.

28. Mellaart, *Neolithic of the Near East*, 108.

29. Nigel Goring-Morris and Liora Kolska Horwitz, "Funerals and Feasts during the Pre-Pottery Neolithic B of the Near East," *Antiquity* 81 (2007): 902–919.

30. M. A. Littauer and J. H. Crouwel, *Wheeled Vehicles and Ridden Animals in the Ancient Near East* (Leiden/Köln: E. J. Brill, 1979), 13.

31. Sumer was the civilization of southern Iraq that lasted from the late sixth millennium BCE through the Uruk period in the fourth millennium BCE until the rise of Babylon in the second millennium BCE. The Sumerians had cities and yearlong agriculture. Akkadian was the language spoken and written (in cuneiform) by the Sumerians in the north of the region from around 2800 BCE.

32. F. E. Zeuner, *A History of Domesticated Animals* (London: Hutchinson, 1963), 367.

33. Hans-Peter Uerpmann, *The Ancient Distribution of Ungulate Mammals in the Middle East* (Wiesbaden: Dr Reichert, 1987), 30.

34. Carles Vilà, Jennifer A. Leonard, and Albano Beja-Pereira, "Genetic Documentation of Horse and Donkey Domestication," in Zeder et al., *Documenting Domestication*, 350.

35. J. Nicholas Postgate, "The Equids of Sumer Again," in *Equids in the Ancient World*, vol. 1, ed. by R. H. Meadow and H.-P. Uerpmann (Wiesbaden: Dr Reichert, 1986), 194.

36. Postgate, "Equids," 1:202.

37. Juliet Clutton-Brock, "Ritual Burial of a Dog and Six Domestic Donkeys," in *Excavations at Tell Brak: Nagar in the Third Millennium b.c.*, vol. 2, ed. by D. Oates, J. Oates, and H. McDonald (Cambridge: British School of Archaeology in Iraq and the McDonald Institute for Archaeological Research, 2001), 327–338.

38. Vilà, Leonard, and Beja-Pereira, "Genetic Documentation," 342–348.

39. The Chalcolithic period is the name given to the period between the Neolithic and the Bronze Age, from around 4500 to 3500 BCE.

40. Uerpmann, *Ancient Distribution of Ungulate Mammals*, 16.

41. Joan Oates, "A Note on the Early Evidence for Horse and the Riding of Equids in Western Asia," in *Prehistoric Steppe Adaptation and the Horse*, ed. by Marsha Levine, Colin Renfrew, and Katie Boyle (Cambridge: McDonald Institute for Archaeological Research, 2003), 115–125.

CHAPTER 3. ARRIVAL OF DOMESTICATES IN EUROPE

1. V. Gordon Childe, *The Dawn of European Civilization* (London: Kegan Paul, 1925).

2. V. Gordon Childe, *The Prehistory of European Society* (London: Penguin Books, 1958), 8.

3. Jean-Denis Vigne, "Zooarchaeological Aspects of the Neolithic Diet Transition in the Near East and Europe and Their Putative Relationships with the Neolithic Demographic Transition," in *The Neolithic Demographic Transition and Its Consequences*, ed. by J.-P. Bocquet Appel and O. Bar-Yosef (New York: Springer Verlag, 2008), 179–204.

4. A. J. Ammerman and L. L. Cavalli-Sforza, "Measuring the Rate of Spread of Early Farming in Europe," *Man* 6 (1971): 674–688.

5. Vigne, "Zooarchaeological Aspects," 191.

6. Judith E. King, "Mammal Bones from Khirokitia and Erimi," in *Khirokitia: Final Report on the Excavation of a Neolithic Settlement in Cyprus on Behalf of the Department of Antiquities, 1936–1946*, by P. Dikaios (Oxford: Oxford University Press, 1953), 431–437 (appendix 111).

7. Alan H. Simmons, *Faunal Extinction in an Island Society: Pygmy Hippopotamus Hunters of Cyprus* (New York: Springer, 1999).

8. Liora Kolska Horwitz and Gila Kahila Bar-Gal, "The Origin and Genetic Status of Insular Caprines in the Eastern Mediterranean: A Case Study of Free-Ranging Goats (*Capra aegagrus cretica*) on Crete," *Human Evolution* 21 (2006): 124.

9. F. Poplin, "Origine du mouflon de Corse dans une nouvelle perspective paléontologique: par marronage," *Annales Génétique et de Sélection Animale* 11, no. 2 (1979): 133–143.

10. Horwitz and Bar-Gal, "Origin and Genetic Status of Insular Caprines," 128.

11. J. Clutton-Brock, K. Dennis-Bryan, P. L. Armitage, and P. A. Jewell, "Osteology of the Soay Sheep," *Bulletin British Museum* 55, no. 2 (1990): 1–56; Tim Clutton-Brock and Josephine Pemberton, *Soay Sheep Dynamics and Selection in an Island Population* (Cambridge: Cambridge University Press, 2004).

12. Paul Halstead, "Sheep in the Garden: The Integration of Crop and Livestock Husbandry in Early Farming Regimes of Greece and Southern Europe," in *Animals in the Neolithic of Britain and Europe*, ed. by Dale Serjeantson and David Field (Oxford: Oxbow Books, 2006), 42–55.

13. Daniel G. Bradley and David A. Magee, "Genetics and the Origins of Domestic Cattle," in *Documenting Domestication: New Genetic and Archaeological Paradigms*, ed. by Melinda A. Zeder, Daniel G. Bradley, Eve Emshwiller, and Bruce D. Smith (Berkeley: University of California Press, 2006), 326.

14. Michael W. Bruford and Saffron J. Townsend, "Mitochondrial DNA Diversity in Modern Sheep," in Zeder et al., *Documenting Domestication*, 306–316.

15. G. Luikart, H. Fernandez, M. Mashkour, P. R. England, and P. Taberlet, "Origins and Diffusion of Domestic Goats Inferred from DNA Markers: Example Analyses of mtDNA, Y Chromosome, and Microsatellites," in Zeder et al., *Documenting Domestication*, 294–305.

16. Luikart, Fernandez, Mashkour, England, and Taberlet, "Origins and Diffusion of Domestic Goats Inferred from DNA Markers," 303.

17. Greger Larson, Umberto Albarella, Keith Dobney, and Peter Rowley-Conwy, "Current Views on *Sus* Phylogeography and Pig Domestication as Seen through Modern mtDNA Studies," in *Pigs and Humans: 10,000 Years of Interaction*, ed. by Umberto Albarella, Keith Dobney, Anton Ervynck, and Peter Rowley-Conwy (Oxford: Oxford University Press, 2007), 30–41.

18. Robin Lane Fox, *Travelling Heroes: Greeks and Their Myths in the Epic Age of Homer* (London: Allen Lane, 2008), 381.

19. Homer, *Odyssey*, trans. by E. V. Rieu (London: Penguin Books, 1946), 215.

20. Thomas Jansen et al., "Mitochondrial DNA and the Origins of the Domestic Horse," *Proceedings of the National Academy of Sciences* 99, no. 16 (August 2002), 10905–10910.

21. Sandra L. Olsen, "Early Horse Domestication on the Eurasian Steppe," in Zeder et al., *Documenting Domestication*, 251–253.

22. Carles Vilà, Jennifer A. Leonard, and Albano Beja-Pereira, "Genetic Documentation of Horse and Donkey Domestication," in Zeder et al., *Documenting Domestication*, 342.

23. Vilà, Leonard, and Beja-Pereira, "Genetic Documentation of Horse and Donkey Domestication," 346.

24. David R. Harris, "The Origins and Spread of Agriculture and Pastoralism in Eurasia: An Overview," in *The Origins and Spread of Agriculture and Pastoralism in Eurasia*, ed. by David R. Harris (London: UCL Press, 1996), 560–561.

25. Mark N. Cohen, *The Food Crisis in Prehistory* (New Haven, Conn.: Yale University Press, 1977).

26. Andrew Sherratt, "Plough and Pastoralism: Aspects of the Secondary Products Revolution," in *Pattern of the Past: Studies in Honour of David Clarke*, ed. by I. Hodder, G. Isaac, and N. Hammond. (Cambridge: Cambridge University Press, 1981), 261–305.

27. Richard P. Evershed et al., "Earliest Date for Milk Use in the Near East and Southeastern Europe Linked to Cattle Herding," *Nature* 455 (25 September 2008): 528–531.

28. E. C. Amoroso and P. A. Jewell, "The Exploitation of the Milk-Ejection Reflex by Primitive Peoples," in *Man and Cattle*, ed. by A. E. Mourant and F. E. Zeuner (Royal Anthropological Institute Occasional Paper 18, 1963), 126–138.

29. Alan K. Outram et al., "The Earliest Horse Harnessing and Milking," *Science* 323, no. 5919, (6 March 2009): 1332–1335.

30. Yuval Itan, Adam Powell, Mark A. Beaumont, Joachim Burger, and Mark G. Thomas, "The Origins of Lactase Persistence in Europe," *PLoS Computational Biology* 5, no. 8 (2009): 1–13.

31. Stuart Piggott, *The Earliest Wheeled Transport from the Atlantic Coast to the Caspian Sea* (London: Thames and Hudson, 1983).

32. Piggott, *The Earliest Wheeled Transport*, 60.

33. Piggott, *The Earliest Wheeled Transport*, 89.

34. M. A. Littauer and J. H. Crouwel, *Wheeled Vehicles and Ridden Animals in the Ancient Near East* (Leiden/Köln: E. J. Brill, 1979), 68.

35. R. L. Fitts, C. Haselgrove, P. Lowther, and S. Willis, "Melsonby Revisited: Survey and Excavation 1992–95 at the Site of Discovery of the 'Stanwick' North Yorkshire Hoard of 1843," *Durham Archaeological Journal* 14–15 (1999): 1–52.

36. A. M. Khazanov, *Nomads and the Outside World*, trans. by J. Crookenden (Cambridge: Cambridge University Press, 1984), 18–20.

37. That the Przewalski horse was never ancestral to domestic horses is proved by mtDNA analysis that has shown these horses to have four quite separate lineages from all other horse breeds. See Jansen et al., "Mitochondrial DNA," 10905–10910.

38. Sevyan Vainshtein, *Nomads of South Siberia: The Pastoral Economies of Tuva*, trans. by Michael Colenso (Cambridge: Cambridge University Press, 1980), 51.

39. Vainshtein, *Nomads of South Siberia*, 51–53.

40. Khazanov, *Nomads and the Outside World*, 30.

CHAPTER 4. DOMESTICATES IN ANCIENT EGYPT AND THEIR ORIGINS

1. A. Muzzolini, "The Emergence of a Food-Producing Economy in the Sahara," in *The Archaeology of Africa: Food, Metals and Towns*, ed. by Thurstan Shaw, Paul Sinclair, Bassey Andah, and Alex Okpoko (New York: Routledge, 1993), 227–229.
2. The chronology of ancient Egypt used in this chapter is from Zahi Hawass, http://www.guardians .net/hawsaa/chronology.htm (20 March 2011).
3. Muzzolini, "The Emergence of a Food-Producing Economy in the Sahara," 229–231.
4. Muzzolini, "The Emergence of a Food-Producing Economy in the Sahara," 233–234.
5. Savino Di Lernia and Mauro Crernasch, "Taming Barbary Sheep: Wild Animal Management by Early Holocene Hunter-Gatherers at Uan Afuda (Libyan Sahara)," *Society of African Archaeologists: Nyame Akuma* 46 (December 1996): 43–54.
6. Carlos Driscoll, Juliet Clutton-Brock, Andrew C. Kitchener, and Stephen J. O'Brien, "The Taming of the Cat," *Scientific American* 300, no. 6 (2009): 56–63.
7. Colin P. Groves, "The Taxonomy, Distribution, and Adaptation of Recent Equids," in *Equids in the Ancient World*, ed. by Richard H. Meadow and Hans-Peter Uerpmann (Wiesbaden: Dr Luwig Reichart, 1986), 27–32.
8. Carles Vilà, Jennifer A. Leonard, and Albano Beja-Pereira, "Genetic Documentation of Horse and Donkey Domestication," in *Documenting Domestication: New Genetic and Archaeological Paradigms*, ed. by Melinda A. Zeder, Daniel G. Bradley, Eve Emshwiller, and Bruce D. Smith (Berkeley: University of California Press, 2006), 348–350.
9. Groves, "Taxonomy," 33.
10. Daniel G. Bradley and David A. Magee, "Genetics and the Origins of Domestic Cattle," in Zeder et al., *Documenting Domestication*, 317–321.
11. Caroline Grigson, "The Craniology and Relationships of Four Species of *Bos. 5. Bos indicus* L.," *Journal of Archaeological Science* 7 (1980): 3–32.
12. Bradley and Magee, "Genetics," 320, 326.
13. Fred Wendorf and Romuald Schild, "Nabta Playa and Its Role in Northeast African Prehistory," *Journal of Anthropological Archaeology* 17 (1998): 97–123.
14. Herodotus, *The Histories of Herodotus*, vol. 1, trans. by George Rawlinson (London: Dent, Everyman's Library, 1970), 358.
15. Herodotus, *The Histories of Herodotus*, 1:360.
16. Herodotus, *The Histories of Herodotus*, 1:132–153.
17. Diodorus Siculus, *Diodorus Siculus*, vol. 1, trans. by C. H. Oldfather (Cambridge, Mass.: Harvard University Press, 1989), 283.
18. Siculus, *Diodorus Siculus*, 1:119.
19. Siculus, *Diodorus Siculus*, 1:303.
20. Patrick F. Houlihan, *The Animal World of the Pharaohs* (New York: Thames and Hudson, 1996), 22.
21. Herodotus, *Histories*, 1:137–138.
22. Houlihan, *Animal World of the Pharaohs*, 26–29.
23. Douglas J. Brewer, Donald B. Redford, and Susan Redford, *Domestic Plants and Animals: The Egyptian Origins* (Warminster, England: Aris and Phillips, 1996), 104–105.
24. Herodotus, *Histories*, 1:362.
25. Vilà, Leonard, and Beja-Pereira, "Genetic Documentation," 348–349, figs. 24.4–24.5.
26. Houlihan, *Animal World of the Pharaohs*, 29–33.

27. Peter Raulwing and Juliet Clutton-Brock, "The Buhen Horse: Fifty Years after Its Discovery (1958–2008)," *Journal of Egyptian History* 2, nos. 1–2 (2009): 1–106.

28. M. A. Littauer and J. H. Crouwel, *Wheeled Vehicles and Ridden Animals in the Ancient Near East* (Leiden/Köln: E. J. Brill, 1979), 75–80, figs. 43–45.

29. Houlihan, *Animal World of the Pharaohs*, 78.

30. Siculus, *Diodorus Siculus*, 1:viii, 285–287.

31. Juliet Clutton-Brock, *The British Museum Book of Cats Ancient and Modern* (London: British Museum Press and the Natural History Museum, 1988), 36–38.

32. Patrick F. Houlihan, *The Birds of Ancient Egypt* (Warminster, England: Aris and Phillips, 1986).

33. Houlihan, *The Birds of Ancient Egypt*, 102, fig. 148.

34. Siculus, *Diodorus Siculus*, 1:257.

35. Eva Crane, "Honeybees," in *Evolution of Domesticated Animals*, ed. by I. L. Mason (New York: Longman, 1984), 403–415.

36. L. Garnery, J.-M. Cornuet, and M. Solignac, "Evolutionary History of the Honey Bee *Apis mellifera* Inferred from Mitochondrial DNA Analysis," *Molecular Ecology* 1, no. 3 (28 June 2008): 145–154.

37. Houlihan, *Animal World of the Pharaohs*, 189.

CHAPTER 5. DOMESTICATES OF THE ANCIENT ISRAELITES, ASSYRIANS, AND SCYTHIANS

1. Peter Parr, "The Levant in the Early First Millennium B.C.," in *The Cambridge Encyclopedia of Archaeology*, ed. by Andrew Sherratt and Grahame Clark (Cambridge: Cambridge University Press, 1980), 197.

2. H. B. Tristram, *The Natural History of the Bible* (London: Society for Promoting Christian Knowledge, 1889).

3. F. S. Bodenheimer, *Animal and Man in Bible Lands* (Leiden: E. J. Brill, 1960); James George Frazer, *The Golden Bough: A Study in Magic and Religion*, 2 vols. (London: Macmillan, 1911–1915).

4. Job 30:1.

5. Job 39:21–25.

6. 1 Samuel 25:20.

7. Judges 5:10.

8. Tristram, *Natural History of the Bible*, 39.

9. Deuteronomy 22:10.

10. Job 39:5–8.

11. Genesis 32:13–15.

12. 2 Chronicles 9:1.

13. Isaiah 21: 6–7.

14. Tristram, *Natural History of the Bible*, 68.

15. 1 Samuel 15:14–15.

16. A. T. Clason, "Late Bronze Age—Iron Age Zebu Cattle in Jordan?" *Journal of Archaeological Science* 5 (1978): 91–93.

17. Genesis 27:9.

18. Joshua 9:4.

19. Tristram, *Natural History of the Bible*, 135.

20. Although detailed molecular analysis has revealed much about breeds of European sheep, only preliminary genetic studies have been carried out on the three groups of sheep that are found in the arid regions of the Middle East and Africa—these being fat-tailed, thin-tailed, and fat-rumped sheep—and no definitive results have so far been published on their origins or relationships.

21. Leviticus 3:9, 11.

22. Leviticus 7:23.

23. Leviticus 11:7, 8; Deuteronomy 14:8.

24. Qur'an 5:4.

25. 2 Kings 19:32, 35.

26. M. A. Littauer and J. H. Crouwel, *Wheeled Vehicles and Ridden Animals in the Ancient Near East* (Leiden/Köln: E. J. Brill, 1979).

27. Herodotus, *The Histories of Herodotus*, vol. 1, trans. by George Rawlinson (London: Dent, Everyman's Library, 1970), 305.

28. Herodotus, *The Histories of Herodotus*, 1:310.

29. A felly is the rim of a wheel.

30. Herodotus, *Histories*, 1:315.

31. S. I. Rudenko, *Frozen Tombs of Siberia: The Pazyryk Burials of Iron-Age Horsemen* (London: Dent, 1970).

32. Herodotus, *Histories*, 1:294.

33. Sevyan Vainshtein, *Nomads of South Siberia: The Pastoral Economies of Tuva*, trans. by Michael Colenso (Cambridge: Cambridge University Press, 1980).

34. Marco Polo, *The Travels of Marco Polo*, trans. by Ronald Latham (London: Penguin Books, 1988), 108–109.

35. Polo, *The Travels of Marco Polo*, 150–152.

36. Polo, *The Travels of Marco Polo*, 330.

CHAPTER 6. DOMESTICATES IN THE CLASSICAL WORLD OF GREECE AND ROME

1. The papers and records of the excavations of Knossos by Sir Arthur Evans are archived at the Ashmolean Museum, Oxford.

2. Michael Rice, *The Power of the Bull* (London: Routledge, 1998), 202–211.

3. Homer, *Iliad*, trans. by Martin Hammond (London: Penguin Books, 1987); Homer, *Odyssey*, trans. by E. V. Rieu (London: Penguin Books, 1946).

4. Homer, *Iliad*, 98.

5. Homer, *Iliad*, 105.

6. Homer, *Iliad*, 379.

7. Homer, *Iliad*, 378.

8. Homer, *Odyssey*, 80.

9. Homer, *Odyssey*, 266–267.

10. Homer, *Odyssey*, 267.

11. Aesop, *Fables of Aesop*, trans. by S. A. Handford (London: Penguin Classics, 1954), no. 96, 100.

12. Aristotle, *Aristotle: Historia Animalium*, Loeb Classical Library, books ɪ–ɪɪɪ, trans. by A. L. Peck

(Cambridge, Mass.: Harvard University Press, 1965); Aristotle, *Aristotle: Historia Animalium*, Loeb Classical Library, books IV–VI, trans. by A. L. Peck (Cambridge, Mass.: Harvard University Press, 1970); Aristotle, *Aristotle: Historia Animalium*, Loeb Classical Library, books VII–X, trans. by D. M. Balme (Cambridge, Mass.: Harvard University Press, 1991).

13. Aristotle, *Aristotle: Generation of Animals*, Loeb Classical Library, trans. by A. L. Peck (Cambridge, Mass.: Harvard University Press, 1990).

14. Aristotle, *Aristotle: Parts of Animals, Movement of Animals, Progression of Animals*, Loeb Classical Library, trans. by A. L. Peck and E. S. Forster (Cambridge, Mass.: Harvard University Press, 1983).

15. Aristotle, *Aristotle: Historia Animalium*, book VII(VIII) (p. 199).

16. Arrian, *Arrian: History of Alexander and Indica*, vol. 1, Loeb Classical Library, trans. by P. A. Brunt (Cambridge, Mass.: Harvard University Press, 1976); Arrian, *Arrian: History of Alexander and Indica*, vol. 2, trans. by P. A. Brunt (Cambridge, Mass.: Harvard University Press, 1983).

17. Arthur Hugh Clough, ed., *Plutarch's Lives: The Dryden Translation* (New York: Modern Library, 2001), 2:143.

18. Arrian, *Arrian: History of Alexander and Indica*, Loeb Classical Library, vol. 2 (pp. 43, 47, 61).

19. Xenophon, *The Persian Expedition*, trans. by Rex Warner (London: Penguin Books, 1975).

20. Xenophon, "On Horsemanship," in *Xenophon's Minor Works*, trans. by J. S. Watson (London: George Bell, 1884); Xenophon, "On the Duties of a Commander of Cavalry," in *Xenophon's Minor Works*.

21. Xenophon, "On Horsemanship," 301.

22. Xenophon, "On the Duties of a Commander," 312.

23. Xenophon, "On Hunting," in *Xenophon's Minor Works*, 330–371.

24. Xenophon, "On Hunting," 354.

25. Clough, *Plutarch's Lives*, 1:31–34.

26. M. P. Cato and M. T. Varro, *Marcus Porcius Cato on Agriculture, Marcus Terentius Varro on Agriculture*, Loeb Classical Library, trans. by W. D. Hooper and H. B. Ash (Cambridge, Mass.: Harvard University Press, 1967), Varro, 423–425.

27. The wars from 264 to 241 BCE between Rome and the Carthaginians from Tunisia were known to the Romans as the Punic Wars because Punici was the Latin name for the Carthaginians.

28. Livy, *The War with Hannibal*, trans. by Aubrey de Sélincourt (London: Penguin Books, 1965), 50–53.

29. Polybius, *The Rise of the Roman Empire*, trans. by Ian Scott-Kilvert (London: Penguin Books, 1979), 429.

30. Cato and Varro, *Marcus Cato*, ix–xiii.

31. Cato and Varro, *Marcus Cato*, LVII, lix. 1 (73).

32. Cato and Varro, *Marcus Terentius Varro*, xvii.

33. Cato and Varro, *Marcus Terentius Varro*, III, xii, 6–7 (493).

34. Cato and Varro, *Marcus Terentius Varro*, III, vii, 1–3 (461–463).

35. L. J. M. Columella, *Lucius Junius Moderatus Columella on Agriculture: De Re Rustica*, vol. 2, Loeb Classical Library, trans. by E. S. Forster and E. H. Heffner (Cambridge, Mass.: Harvard University Press, 1968), books V–IX.

36. Columella, *Lucius Junius Moderatus Columella on Agriculture*, vol. 2 (p. 315).

37. Columella, *Lucius Junius Moderatus Columella on Agriculture*, vol. 2 (p. 125).

38. Columella, *Lucius Junius Moderatus Columella on Agriculture*, vol. 2 (p. 235).

39. Columella, *Lucius Junius Moderatus Columella on Agriculture*, vol. 2 (pp. 323–327).

40. Pliny, *The Natural History of Pliny*, 6 vols., trans. by J. Bostock (London: Henry G. Bohn, 1855).

41. Pliny, *The Natural History of Pliny*, vol. 2 (p. 350).

42. Clyde Pharr, trans., *The Theodosian Code and Novels and the Sirmondian Constitutions* (Princeton, N.J.: Princeton University Press, 1952).

43. One Roman pound (libra) was equivalent to approximately 327 grams, or .75 of an English pound.

44. Pharr, *Theodosian Code*, Title 31. 8.1–2 (196).

45. Pharr, *Theodosian Code*, Title 17.3 (310).

46. Pharr, *Theodosian Code*, Title 10.1 (436).

CHAPTER 7. DOMESTICATES IN ANCIENT INDIA AND SOUTHEAST ASIA

1. Hasmukh Dhirajlal Sankalia, *Excavations at Langhnaj: 1944–63 Part I: Archaeology* (Poona: Deccan College Postgraduate and Research Institute, 1965); Juliet Clutton-Brock, *Excavations at Langhnaj: 1944–63 Part II: The Fauna* (Poona: Deccan College Postgraduate and Research Institute, 1965).

2. Richard H. Meadow, "The Origins and Spread of Agriculture and Pastoralism in Northwestern South Asia," in *The Origins and Spread of Agriculture and Pastoralism in Eurasia*, ed. by David R. Harris (London: UCL Press, 1996), 402–406.

3. Meadow, "The Origins and Spread of Agriculture and Pastoralism," 405.

4. Richard H. Meadow, personal communication with author, April 2010; Bridget Allchin and Raymond Allchin, *The Rise of Civilization in India and Pakistan* (Cambridge: Cambridge University Press, 1982), 166–171.

5. Allchin and Allchin, *Rise of Civilization*, 178–186.

6. Meadow, personal communication.

7. Seema J. Parwanker, *Man and Animal Relationships in Early Farming Communities of Western India with Special Reference to Inamgaon* (Oxford: BAR, S, 1639, 2007).

8. M. K. Dhavalikar, "From Farming to Pastoralism: Effects of Climatic Change in the Deccan," in *The Walking Larder: Patterns of Domestication, Pastoralism and Predation*, ed. by Juliet Clutton-Brock (London: Unwin Hyman, 1989), 156–168.

9. Juliet Clutton-Brock, *A Natural History of Domesticated Mammals*, 2nd ed. (Cambridge: Cambridge University Press, 1999), 170–172; J. Bonnemaire, "Yak," in *Evolution of Domesticated Animals*, ed. by I. L. Mason (New York: Longman, 1984), 9–45.

10. Caroline Grigson, "The Craniology and Relationships of Four Species of *Bos* 5. *Bos indicus* L." *Journal of Archaeological Science* 7 (1980): 3–32.

11. R. T. Loftus, D. E. Machugh, D. G. Bradley, P. M. Sharp, and P. Cunningham, "Evidence for Two Independent Domestications of Cattle," *Proceedings of the National Academy of Sciences*, 91 (1994): 2757–2761.

12. J. S. F. Barker, S. S. Moore, D. J. S. Hettzel, D. Evans, S. G. Tan, and K. Byrne, "Genetic Diversity of Asian Water Buffalo (*Bubalus bubalis*): Microsatellite Variation and a Comparison with Protein-Coding Loci," *Animal Genetics* 28, no. 2 (1997): 103–115.

13. F. J. Simoons, "Gayal or Mithan," in Mason, *Evolution of Domesticated Animals*, 34–39; Clutton-Brock, *Natural History of Domesticated Mammals*, 167–170.

14. Arrian, *Arrian: History of Alexander and Indica*, vol. 2, Loeb Classical Library, trans. by P. A. Brunt (Cambridge, Mass.: Harvard University Press, 1983), 27–59.

15. F. E. Zeuner, *A History of Domesticated Animals* (London: Hutchinson, 1963), 292, 294.

16. Simon Hedges and Donny Gunaryadi, "Reducing Human-Elephant Conflict: Do Chillies Help Deter Elephants from Entering Crop Fields?" *Oryx* 44, no. 1 (2009): 133–138.

17. Greger Larson, Umberto Albarella, Keith Dobney, and Peter Rowley-Conwy, "Current Views on *Sus* Phylogeography and Pig Domestication as Seen through Modern mtDNA Studies," in *Pigs and Humans: 10,000 Years of Interaction*, ed. by Umberto Albarella, Keith Dobney, Anton Ervynck, and Peter Rowley-Conwy (Oxford: Oxford University Press, 2007), 30–41.

18. Greger Larson et al., "Worldwide Phylogeography of Wild Boar Reveals Multiple Centres of Pig Domestication," *Science* 307 (2005): 1618–1621.

19. Colin Groves, "Current Views on Taxonomy and Zoogeography of the Genus *Sus*," in Albarella et al., *Pigs and Humans*, 18–29.

20. H. Epstein, *Domestic Animals of China* (Bucks, England: Commonwealth Agricultural Bureaux, Farnham Royal, 1969), 69–93.

21. Yong-fu Huang, Xian-wei Shi, and Ya-ping Zhang, "Mitochondrial Genetic Variation in Chinese Pigs and Wild Boars," *Biochemical Genetics* 37, nos. 11/12 (1999): 335–343.

22. Greger Larson et al., "Patterns of East Asian Pig Domestication, Migration, and Turnover Revealed by Modern and Ancient DNA," *Proceedings of the National Academy of Sciences* 107, no. 17 (27 April 2010): 7686–7691.

23. Helmut Hemmer, *Domestication: The Decline of Environmental Appreciation* (Cambridge: Cambridge University Press, 1990), 40–43.

24. Peter Savolainen, Ya-ping Zhang, Jing Luo, Joakim Lundeberg, and Thomas Leitner, "Genetic Evidence for an East Asian Origin of Domestic Dogs," *Science* 298 (22 November 2002): 1610–1613.

25. Jun-Feng Pang et al. "mtDNA Data Indicate a Single Origin for Dogs South of Yangtze River, Less than 16,300 Years Ago, from Numerous Wolves," *Molecular Biology and Evolution* 26, no. 12 (2009): 2849–2864.

26. Epstein, *Domestic Animals of China*, 126.

27. Anthony L. Podberscek, "Good to Pet and Eat: The Keeping and Consuming of Dogs and Cats in South Korea," *Journal of Social Issues* 65, no. 3 (2009): 617.

28. Epstein, *Domestic Animals of China*, 136.

29. E. C. Ash, *Dogs: Their History and Development* (New York: Benjamin Blom, 1972), 2:619–621.

30. Epstein, *Domestic Animals of China*, 131.

31. Marco Polo, *The Travels of Marco Polo*, trans. by Ronald Latham (London: Penguin Books, 1988).

32. Epstein, *Domestic Animals of China*, 145.

33. Naotaka Ishiguro, Naohiko Okumura, Akira Matsui, and Nobuo Shigehara, "Molecular Genetic Analysis of Ancient Japanese Dogs," in *Dogs through Time: An Archaeological Perspective*, ed. by Susan Crockford (Oxford: BAR International Series 889, 2000), 287–292.

34. Ash, *Dogs*, 2: 624–629.

35. Carlos A. Driscoll et al., "Near Eastern Origin of Cat Domestication," *Science* 317 (2007): 519–523.

36. Monika J. Lipinski et al., "The Ascent of Cat Breeds: Genetic Evaluations of Breeds and Worldwide Random Bred Populations," *Genomics* 91, no. 1 (2008): 12–21.

37. R. D. Crawford, "Domestic Fowl," in Mason, *Evolution of Domesticated Animals*, 298–311.

38. J. Eriksson et al., "Identification of the *Yellow Skin* Gene Reveals a Hybrid Origin of the Domestic Chicken," *PLoS Genetics* 4, no. 2: e1000010, 29 February 2008, doi:10.1371/journal.pgen.1000010.

39. Barbara West and Ben-Xiong Zhou, "Did Chickens Go North? New Evidence for Domestication," *World's Poultry Science Journal* 45 (1989): 205–218.

40. M. P. Cato and M. T. Varro, *Marcus Porcius Cato on Agriculture, Marcus Terentius Varro on Agriculture*, Loeb Classical Library, trans. by W. D. Hooper and H. B. Ash (Cambridge, Mass.: Harvard University Press, 1967), Cato, 91.

41. Varro, 471–483.

42. L. J. M. Columella, *Lucius Junius Moderatus Columella on Agriculture: De Re Rustica*, vol. 2, Loeb Classical Library, trans. by E. S. Forster and E. H. Heffner (Cambridge, Mass.: Harvard University Press, 1968), 323–361.

43. Columella, *Lucius Junius Moderatus Columella on Agriculture*, vol. 2, 325.

44. Iain Grahame, "Peafowl," in Mason, *Evolution of Domesticated Animals*, 315–321.

45. 1 Kings 10:22.

46. Diodorus Siculus, *Diodorus Siculus*, Loeb Classical Library, trans. by C. H. Oldfather (Cambridge, Mass.: Harvard University Press, 1989), book II, 53, 2 (vol. II, p. 59).

47. Cato and Varro, *Marcus Terentius Varro on Agriculture*, book III, VI, 1 (p. 459).

48. Columella, *Lucius Junius Moderatus Columella on Agriculture*, book VIII, XI, 1–15 (vol. II, pp. 375–385).

49. Zhang Zhon-ge, "Goldfish," in Mason, *Evolution of Domesticated Animals*, 381–385.

50. Tomoyoshi Komiyama, Hiroyuki Kobayashia, Yoshio Tatenob, Hidetoshi Inokoa, Takashi Gojoborib, and Kazuho Ikeob, "An Evolutionary Origin and Selection Process of Goldfish," *Gene* 430, nos. 1–2 (1 February 2009): 5–11.

51. Zhon-ge, "Goldfish," 382.

52. Zhon-ge, "Goldfish," 383.

CHAPTER 8. DOMESTICATES IN OCEANIA

1. Manfred Kayser, "The Human Genetic History of Oceania: Near and Remote Views of Dispersal," *Current Biology* 20, no. 4 (February 2010): 202–207.

2. M. J. Morwood et al., "Archaeology and Age of a New Hominin from Flores in Eastern Indonesia," *Nature* 431 (28 October 2004): 1087–1091.

3. Susan Bulmer, "Lapita Dogs and Singing Dogs and the History of the Dog in New Guinea," in *The Archaeology of Lapita Dispersal in Oceania: Papers from the Fourth Lapita Conference, Canberra*, ed. by G. R. Clark, A. J. Anderson, and T. Vunidilo (Canberra: Pandanus Books, 2001), 183–201.

4. Bulmer, "Lapita Dogs and Singing Dogs," 185.

5. Ellis Troughton, "A New Native Dog from the Papuan Highlands," *Proceedings of the Royal Zoological Society of New South Wales, 1955–56* (8 May 1957): 93–94.

6. Janice Koler-Matznick, I. Lehr Brisbin Jr., Mark Feinstein, and Susan Bulmer, "An Updated Description of the New Guinea Singing Dog (*Canis hallstromi*, Troughton 1957)," *Journal of Zoology* 261, no. 2 (2003): 109–118.

7. L. K. Corbett, "Morphological Comparisons of Australian and Thai Dingoes: A Reappraisal of Dingo Status, Distribution and Ancestry," *Proceedings of the Ecological Society of Australia* 13 (1985): 277–290.

8. Peter Savolainen, Thomas Leitner, Alan N. Wilton, Elizabeth Matisoo-Smith, and Joakim Lundeberg, "A Detailed Picture of the Origin of the Australian Dingo, Obtained from the Study of Mitochondrial DNA," *Proceedings of the National Academy of Sciences* 101, no. 33 (17 August 2004): 12387–12390.

9. C. Lumholz, *Among Cannibals* (London: J. Murray, 1899); M. Meggitt, "Australian Aborigines and Dingoes," in *Man, Culture and Animals*, ed. by A. Leeds and P. Vayda (American Association for the Advancement of Science Symposium, 1965), 7–26; Bradley P. Smith and Carla A. Litchfield, "A Review of the Relationship between Indigenous Australians, Dingoes (*Canis dingo*) and Domestic Dogs (*Canis familiaris*)," *Anthrozoös* 22, no. 2 (2009): 111–128.

10. Margaret Titcomb, *Dog and Man in the Ancient Pacific with Special Attention to Hawaii* (Bernice P. Bishop Museum Special Publication 59, Honolulu, Hawaii, 1969), 1.

11. Titcomb, *Dog and Man*, 27, quoted from W. Ellis *Polynesian Researches* (London: Fisher, Jackson, 1839), 2:500–501.

12. Colin Groves, *Ancestors for the Pigs*, Technical Bulletin No. 3, Department of Prehistory, Research School of Pacific Studies, Australian National University, 1981.

13. Colin Groves, "Current Views on Taxonomy and Zoogeography of the Genus *Sus*," in *Pigs and Humans: 10,000 Years of Interaction*, ed. by Umberto Albarella, Keith Dobney, Anton Ervynck, and Peter Rowley-Conwy (Oxford: Oxford University Press, 2007), 18–29.

14. Greger Larson et al., "Phylogeny and Ancient DNA of Sus Provides Insights into Neolithic Expansion in Southeast Asia and Oceania," *Proceedings of the National Academy of Sciences* 104 (2007): 4834–4839.

15. Y. P. Liu et al., "Multiple Maternal Origins of Chickens: Out of the Asian Jungles," *Molecular Phylogenetics and Evolution* 38, no. 1 (2006): 12–19.

16. Larson et al., "Phylogeny," 4838.

17. Rebecca Tamsin Jewell, "Understanding Pacific Feather Artefacts through Drawing" (Ph.D. diss., Royal College of Art, 2004), 98–102.

18. Jewell, "Understanding Pacific Feather Artefacts," 102.

19. Harold B. Carter, ed., *The Sheep and Wool Correspondence of Sir Joseph Banks, 1781–1820* (London: Library Council of New South Wales in association with the British Museum [Natural History], 1979), no. 1174, 431.

20. Carter, *The Sheep and Wool Correspondence of Sir Joseph Banks*, no. 1399, 509.

21. Christopher Lever, *Naturalized Mammals of the World* (London: Longman, 1985).

22. C. A. Tisdell, *Wild Pigs: Environmental Pest or Economic Resource?* (Oxford: Pergamon Press, 1982).

23. A. W. Crosby, *The Columbian Exchange: Biological and Cultural Consequences of 1492* (Westport, Conn.: Greenwood Press, 1972), 2–3.

CHAPTER 9. DOMESTICATES IN AFRICA SOUTH OF THE SAHARA

1. Stephen C. Schuster et al., "Complete Khoisan and Bantu Genomes from Southern Africa," *Nature* 463 (18 February 2010): 943–947.

2. David Phillipson, *African Archaeology*, 2nd ed. (Cambridge: Cambridge University Press, 1993), 7.

3. Peter Robertshaw, "The Beginnings of Food Production in Southwestern Kenya," in *The Archaeology of Africa Food, Metals and Towns*, ed. by Thurstan Shaw, Paul Sinclair, Bassey Andah, and Alex Okpoko (London: Routledge, 1993), 359–362.

4. E. E. Evans-Pritchard, *The Nuer* (Oxford: Oxford University Press, 1940).

5. The earliest African cattle are now considered to have been derived from wild *Bos primigenius* in North Africa, with later imports from western Asia. R. T. Loftus, D. E. Machugh, D. G. Bradley,

P. M. Sharp, and P. Cunningham, "Evidence for Two Independent Domestications of Cattle," *Proceedings of the National Academy of Sciences* 91 (1994): 2757–2761.

6. The term "indigenous" is used in this chapter for domesticates that became established in Africa before colonial times or are descended from wild African progenitors, of which there are only two definite species, the donkey and the Guinea fowl, with the cat as a third possibility.

7. Ronan Loftus and Patrick Cunningham, "Molecular Genetic Analysis of African Zeboid Populations," in *The Origins and Development of African Livestock: Genetics, Linguistics and Ethnography*, ed. by Roger M. Blench and Kevin C. MacDonald (London: UCL Press, 2000), 252.

8. H. Epstein, *The Origin of the Domestic Animals of Africa*, 2 vols. (New York: Africana Publishing, 1971), 1:505–535.

9. Loftus and Cunningham, "Molecular Genetic Analysis," 251–252.

10. Ciaran Meghen, David MacHugh, B. Sauveroche, G. Kana, and Dan Bradley, "Characterization of the Kuri Cattle of Lake Chad Using Molecular Genetic Techniques," in Blench and MacDonald, *Origins and Development of African Livestock*, 259–268.

11. Andrew B. Smith, *Pastoralism in Africa: Origins and Development Ecology* (London: Hurst, 1992), 152–155.

12. V. Porter, *Cattle: A Handbook to the Breeds of the World* (London: Christopher Helm, 1991), 221.

13. Kathleen Ryan, Karega Mumene, Samuel M. Kahinju, and Paul N. Kunoni, "Ethnographic Perspectives on Cattle Management in Semi-arid Environments: A Case Study from Maasailand," in Blench and MacDonald, *Origins and Development of African Livestock*, 462–477.

14. Sarah A. Tishkoff et al., "Convergent Adaptation of Human Lactase Persistence in Africa and Europe," *Nature Genetics* 39 (2007): 31–40.

15. D. N. Beach, *The Shona and Zimbabwe 900–1850* (London: Heinemann, 1980), 41–51.

16. Porter, *Cattle*, 229.

17. Richard Klein, "The Prehistory of Stone Age Herders in the Cape Province of South Africa," in *Prehistoric Pastoralism in Southern Africa*, ed. by Martin Hall and Andrew B. Smith (South African Archaeological Society, Goodwin Series, vol. 5, June 1986), 5–12.

18. Andrew B. Smith, "The Origins of the Domesticated Animals of Southern Africa," in Blench and MacDonald, *Origins and Development of African Livestock*, 222–238.

19. Juliet Clutton-Brock, "The Spread of Domestic Animals in Africa," in Shaw et al., *Archaeology of Africa*, 69.

20. Clutton-Brock, "The Spread of Domestic Animals in Africa," 68.

21. "Landrace" is the term used for domesticated animals (or plants) that have evolved in adaptation to the natural and cultural environment in which they live.

22. Clutton-Brock, "Spread of Domestic Animals," 62.

23. Hilde Gauthier-Pilter and Anne Innis Dagg, *The Camel: Its Evolution, Ecology, Behavior and Relationship to Man* (Chicago: University of Chicago Press, 1981), 33.

24. Randolph S. Churchill, *Men, Mines and Animals in South Africa* (London: Sampson Low, Marston, 1891), 191.

25. Roger Blench, "Ethnographic and Linguistic Evidence for the Prehistory of African Ruminant Livestock, Horses and Ponies," in Shaw et al., *Archaeology of Africa*, 89–103.

26. Clutton-Brock, "Spread of Domestic Animals," 64.

27. Johan Gallant, *The Story of the African Dog* (Pietermaritzburg: University of Natal Press, 2002), 4.

CHAPTER 10. DOMESTICATES IN THE AMERICAS

1. Dennis H. O'Rourke and Jennifer A. Raff, "The Human Genetic History of the Americas: The Final Frontier," *Current Biology* 20 (23 February 2010): R202–R207.

2. Colin Renfrew, "Archaeogenetics: Towards a 'New Synthesis,'" *Current Biology* 20 (23 February 2010): R163.

3. Gary Haynes, ed., *American Megafaunal Extinctions at the End of the Pleistocene* (New York: Springer, 2009).

4. Steven Mithen, *After the Ice: A Global Human History, 20,000–5000 b.c.* (London: Weidenfeld and Nicolson, 2003), 246–257.

5. G. M. Allen, "Dogs of American Aborigines," *Bulletin Museum of Comparative Zoology* 63, no. 9 (1920): 431–517.

6. D. Morey and M. D. Wiant, "Early Holocene Domestic Dog Burials from the North American Midwest," *Current Anthropology* 33 (1992): 224–229.

7. J. A. Leonard, Robert K. Wayne, Jane Wheeler, Raúl Valadez, Sonia Guillén, and Carles Vilà, "Ancient DNA Evidence for Old World Origin of New World Dogs," *Science* 298 (2002): 1613–1616; Robert K. Wayne, Jennifer A. Leonard, and Carles Vilà, "Genetic Analysis of Dog Domestication," in *Documenting Domestication: New Genetic and Archaeological Paradigms*, ed. by Melinda A. Zeder et al., (Berkeley: University of California Press, 2006), 284–288.

8. Bridgett M. von Holdt et al., "Genome-Wide SNP and Haplotype Analyses Reveal a Rich History Underlying Dog Domestication," *Nature* 464 (8 April 2010): 898–902.

9. Ben F. Koop and Susan J. Crockford, "Ancient DNA Evidence of a Separate Origin for North American Indigenous Dogs," in *Dogs through Time: An Archaeological Perspective*, ed. by Susan Jane Crockford (Oxford: BAR International Series 889, 2000), 271–284.

10. Marion Schwartz, *A History of Dogs in the Early Americas* (New Haven, Conn.: Yale University Press, 1997).

11. Juliet Clutton-Brock and Norman Hammond, "Hot Dogs: Comestible Canids in Preclassic Maya Culture at Cuello, Belize," *Journal of Archaeological Science* 21 (1994): 819–826.

12. Susan J. Crockford, "A Commentary on Dog Evolution: Regional Variation, Breed Development and Hybridization with Wolves," in Crockford, *Dogs through Time*, 303.

13. Marion Schwartz, *History of Dogs*, 13.

14. R. D. Crawford, "Turkey," in *Evolution of Domesticated Animals*, ed. by I. L. Mason (London: Longman, 1984), 325–334.

15. Crawford, "Turkey," 329.

16. Camilla F. Speller et al., "Ancient Mitochondrial DNA Analysis Reveals Complexity of Indigenous North American Turkey Domestication," *Proceedings of the National Academy of Sciences* 107, no. 7 (16 February 2010): 2807–2812.

17. Crawford, "Turkey," 330.

18. James E. Harting, *The Birds of Shakespeare* (1871; repr., Chicago: Argonaut, 1965), 177.

19. Marion Schwartz, *History of Dogs*, 18.

20. Juliet Clutton-Brock, "The Carnivore Remains Excavated at Fell's Cave in 1970," in *Travels and Archaeology in South Chile by Junius B. Bird*, ed. by J. Hyslop (Iowa City: University of Iowa Press, 1988), 188–195.

21. Jane Wheeler, "On the Origin and Early Development of Camelid Pastoralism in the Andes," in *Animals and Archaeology 3: Early Hunters and Their Flocks*, ed. by J. Clutton-Brock and C. Grigson (Oxford: BAR International Series, 202, 1984), 395–410.

22. G. L. Mengoni Goñalons and H. D. Yacobaccio, "The Domestication of South American Camelids: A View from the South-Central Andes," in Zeder et al., *Documenting Domestication*, 228–244.

23. Miranda Kadwell et al., "Genetic Analysis Reveals the Wild Ancestors of the Llama and the Alpaca," *Proceedings of the Royal Society* B 268 (2001): 2575–2584.

24. J. C. Wheeler, "Evolution and Present Situation of the South American Camelidae," *Biological Journal of the Linnean Society* 54 (1995): 271–295.

25. Kadwell et al., "Genetic Analysis," 2582.

26. J. C. Wheeler, A. J. F. Russel, and H. Redden, "Llamas and Alpacas: Pre-conquest Breeds and Post-conquest Hybrids," *Journal of Archaeological Science* 22, no. 6 (1995): 833–840; Kadwell et al., "Genetic Analysis," 2575.

27. Barbara Weir, "Notes on the Origin of the Domestic Guinea-pig," in *The Biology of Hystricomorph Rodents*, ed. by I. W. Rowlands and Barbara J. Weir (Symposia of the Zoological Society of London, 34, 1974), 437–446; B. Müller-Haye, "Guinea-pig or Cuy," in Mason, *Evolution of Domesticated Animals*, 252–257.

28. Angel E. Spotorno, J. C. Marin, G. Manriquez, J. P. Valladares, E. Rico, and C. Rivas, "Ancient and Modern Steps during the Domestication of Guinea-Pigs," *Journal of Zoology* 270, no. 1 (2006): 57–62.

29. Schwartz, *History of Dogs*, 73.

30. G. A. Clayton, "Muscovy Duck," in Mason, *Evolution of Domesticated Animals*, 340–344.

31. E. Melville, *A Plague of Sheep: Environmental Consequences of the Conquest of Mexico* (Cambridge: Cambridge University Press, 1997), 1.

32. Elinor Melville, *Plague of Sheep*, 6–7.

33. Alfred W. Crosby, *Ecological Imperialism: The Biological Expansion of Europe, 900–1900* (Cambridge: Cambridge University Press, 1986).

34. Melville, *Plague of Sheep*, 51.

35. A. W. Crosby, *The Columbian Exchange: Biological and Cultural Consequences of 1492* (Westport, Conn.: Greenwood Press, 1972).

36. Juliet Clutton-Brock, *Horse Power: A History of the Horse and the Donkey in Human Societies* (London: Natural History Museum Publications, 1992), 144.

37. Clutton-Brock, *Horse Power*, 149.

38. Joel Berger, *Wild Horses of the Great Basin* (Chicago: University of Chicago Press, 1986).

39. Crosby, *Ecological Imperialism*, 177.

40. John J. Mayer and I. Lehr Brisbin Jr., *Wild Pigs of the United States: Their History, Morphology, and Current Status* (Athens: University of Georgia Press, 1991), 7.

41. Crosby, *Ecological Imperialism*, 175.

42. Crosby, *Ecological Imperialism*, 188–189.

CONCLUSIONS

1. Nicholas Guppy, *Wai-Wai through the Forests North of the Amazon* (London: Penguin Books, 1961), 233.

2. The flight distance of a species is the average distance that individual animals within that species will allow between themselves and a potential predator before they flee. Species with a short flight

distance allow potential predators or humans to approach closer than species that have a long flight distance.

3. Elinor Melville, *A Plague of Sheep: Environmental Consequences of the Conquest of Mexico* (Cambridge: Cambridge University Press, 1997), 6–7; Juliet Clutton-Brock, *Horse Power: A History of the Horse and the Donkey in Human Societies* (London: Natural History Museum Publications, 1992), 144.

4. Elizabeth Marshall Thomas, *The Old Way: A Story of the First People* (New York: Picador, 2007).

5. Compassion in World Farming, http://www.ciwf.org.uk.

6. Jonathan Safran Foer, *Eating Animals* (London: Hamish Hamilton, 2009), 129.

7. Charles Kingsley, *The Water Babies* (London: Macmillan, 1910), 203.

APPENDIX. NOMENCLATURE OF THE DOMESTIC ANIMALS AND THEIR WILD PROGENITORS

1. Caroli Linnaei, *Systema Naturae*, facsimile ed. (London: Trustees of the British Museum [Natural History], 1958).

2. International Commission on Zoological Nomenclature, "Usage of 17 Specific Names Based on Wild Species Which Are Predated by or Contemporary with Those Based on Domestic Animals (Lepidoptera, Osteichthyes, Mammalia): Conserved," *Bulletin of Zoological Nomenclature* 60, no. 1 (2003): 81–84.

3. Anthea Gentry, Juliet Clutton-Brock, and Colin P. Groves, "The Naming of Wild Animal Species and Their Domestic Derivatives," *Journal of Archaeological Science* 31 (2004): 645–651.

Bibliography

Aesop. *Fables of Aesop.* Trans. by S. A. Handford. London: Penguin Classics, 1954.

Allchin, Bridget, and Raymond Allchin. *The Rise of Civilization in India and Pakistan.* Cambridge: Cambridge University Press, 1982.

Allen, G. M. "Dogs of American Aborigines." *Bulletin of the Museum of Comparative Zoology* 63, no. 9 (1920): 431–517.

Ammerman, A. J., and L. L. Cavalli-Sforza. "Measuring the Rate of Spread of Early Farming in Europe." *Man* 6 (1971): 674–688.

Amoroso, E. C., and P. A. Jewell. "The Exploitation of the Milk-Ejection Reflex by Primitive Peoples." In *Man and Cattle*, ed. by A. E. Mourant and F. E. Zeuner, 126–138. Royal Anthropological Institute Occasional Paper 18 (1963).

Aristotle. *Aristotle: Generation of Animals*, Loeb Classical Library. Trans. by A. L. Peck. Cambridge, Mass.: Harvard University Press, 1990.

———. *Aristotle: Historia Animalium*, Loeb Classical Library, books 1–3. Trans. by A. L. Peck. Cambridge, Mass.: Harvard University Press, and London: Heinemann, 1965.

———. *Aristotle: Historia Animalium*, Loeb Classical Library, books 4–6. Trans. by A. L. Peck. Cambridge, Mass.: Harvard University Press, 1970.

———. *Aristotle: Historia Animalium*, Loeb Classical Library, books 7–10. Trans. by D. M. Balme. Cambridge, Mass.: Harvard University Press, 1991.

———. *Aristotle: Parts of Animals, Movement of Animals, Progression of Animals*, Loeb Classical Library. Trans. by A. L. Peck and E. S. Forster. Cambridge, Mass.: Harvard University Press, and London: Heinemann, 1983.

Armitage, P. L., and J. Clutton-Brock. "A System for Classification and Description of the Horn Cores of Cattle from Archaeological Sites." *Journal of Archaeological Science* 3 (1976): 329–348.

Arrian. *Arrian: History of Alexander and Indica*, Loeb Classical Library, vol. 1. Trans. by P. A. Brunt. Cambridge, Mass.: Harvard University Press, 1976.

———. *Arrian: History of Alexander and Indica,* Loeb Classical Library, vol. 2. Trans. by P. A. Brunt. Cambridge, Mass.: Harvard University Press, 1983.

Ash, E. C. *Dogs: Their History and Development*, vol. 2. New York: Benjamin Blom, 1972.

Barker, J. S. F., S. S. Moore, D. J. S. Hettzel, D. Evans, S. G. Tan, and K. Byrne. "Genetic Diversity of Asian Water Buffalo (*Bubalus bubalis*): Microsatellite Variation and a Comparison with Protein-Coding Loci." *Animal Genetics* 28, no. 2 (1997): 103–115.

Bar-Yosef, Ofer, and François Valla, eds. *The Natufian Culture in the Levant.* Ann Arbor, Mich.: International Monographs in Prehistory Archaeological Series 1, 1991.

Beach, D. N. *The Shona and Zimbabwe 900–1850.* London: Heinemann, 1980.

Bell, Charles Davidson. *Scraps from My South African Sketch Books.* 1848.

Berger, Joel. *Wild Horses of the Great Basin.* Chicago: University of Chicago Press, 1986.

Belyaev, D., and L. N. Trut. "Some Genetic and Endocrine Effects of Selection for Domestication in Silver Foxes." In *The Wild Canids*, ed. by M. W. Fox. New York: Van Nostrand Reinhold, 1975.

Blench, Roger. "Ethnographic and Linguistic Evidence for the Prehistory of African Ruminant Livestock, Horses and Ponies." In *The Archaeology of Africa: Food, Metals and Towns*, ed. by Thurstan Shaw, Paul Sinclair, Bassey Andah, and Alex Okpoko, 89–103. London: Routledge, 1993.

Bodenheimer, F. S. *Animal and Man in Bible Lands.* Leiden: E. J. Brill, 1960.

Bonnemaire, J. "Yak." In *Evolution of Domesticated Animals*, ed. by I. L. Mason, 9–45. New York: Longman, 1984.

Bradley, Daniel G., and David A. Magee. "Genetics and the Origins of Domestic Cattle." In *Documenting Domestication: New Genetic and Archaeological Paradigms*, ed. by Melinda A. Zeder et al., 317–328. Berkeley: University of California Press, 2006.

Brewer, Douglas J., Donald B. Redford, and Susan Redford. *Domestic Plants and Animals: The Egyptian Origins.* Warminster, England: Aris and Phillips, 1996.

Bruford, Michael W., and Saffron J. Townsend. "Mitochondrial DNA Diversity in Modern Sheep." In *Documenting Domestication: New Genetic and Archaeological Paradigms*, ed. by Melinda A. Zeder et al., 306–316. Berkeley: University of California Press, 2006.

Brunner, Bernd. *The Ocean at Home: An Illustrated History of the Aquarium.* New York: Princeton Architectural Press, 2005.

Budiansky, S. *The Covenant of the Wild: Why Animals Chose Domestication.* New York: William Morrow, 1992.

Bulmer, Susan. "Lapita Dogs and Singing Dogs and the History of the Dog in New Guinea." In *The Archaeology of Lapita Dispersal in Oceania: Papers from the Fourth Lapita Conference, Canberra*, ed. by G. R. Clark, A. J. Anderson, and T. Vunidilo, 183–201. Canberra: Pandanus Books, 2001.

Burleigh, Richard, and Juliet Clutton-Brock. "The Survival of *Myotragus balearicus* Bate, 1909, into the Neolithic on Mallorca." *Journal of Archaeological Science* 7 (1980): 385–388.

Carter, Harold B., ed. *The Sheep and Wool Correspondence of Sir Joseph Banks, 1781–1820.* London: Library Council of New South Wales in association with the British Museum (Natural History), 1979.

Cato, Marcus Porcius, and Marcus Terentius Varro. *Marcus Porcius Cato on Agriculture, Marcus Terentius Varro on Agriculture,* Loeb Classical Library. Trans. by W. D. Hooper and H. B. Ash. Cambridge, Mass.: Harvard University Press, 1967.

Chaix, L., A. Bridault, and R. Picavet. "A Tamed Brown Bear (*Ursus arctos* L.) of the Late Mesolithic from La Grande-Rivoire (Isère, France)?" *Journal of Archaeological Science* 24, no. 12 (1997): 1067–1074.

Childe, V. Gordon. *The Dawn of European Civilization.* London: Kegan Paul, 1925.

———. *The Prehistory of European Society.* London: Penguin Books, 1958.

Churchill, Randolph S. *Men, Mines and Animals in South Africa.* London: Sampson Low, Marston, 1891.

Clark, J. G. D. *Excavations at Star Carr.* Cambridge: Cambridge University Press, 1954.

Clason, A. T. "Late Bronze Age—Iron Age Zebu Cattle in Jordan?" *Journal of Archaeological Science* 5 (1978): 91–93.

Clayton, G. A. "Muscovy Duck." In *Evolution of Domesticated Animals*, ed. by I. L. Mason, 340–344. New York: Longman, 1984.

Clough, Arthur Hugh, ed. *Plutarch's Lives: The Dryden Translation*, vols. 1–2. New York: Modern Library, 2001.

Clutton-Brock, Juliet. "Aristotle, the Scale of Nature, and Modern Attitudes to Animals." In *"Humans and Other Animals*, ed. by Arien Mack, 5–24. Columbus: Ohio State University Press, 1995.

———. *The British Museum Book of Cats Ancient and Modern*. London: British Museum Press and Natural History Museum, 1988.

———. "The Carnivore Remains Excavated at Fell's Cave in 1970." In *Travels and Archaeology in South Chile by Junius B. Bird*, ed. by J. Hyslop, 188–195. Iowa City: University of Iowa Press, 1988.

———. "The Domestication Process: The Wild and Tame." In *Encyclopedia of Human-Animal Relationships*, ed. by Marc Bekoff, 639–643. Westport, Conn.: Greenwood Press, 2007.

———. *Excavations at Langhnaj: 1944–63 Part II: The Fauna*. Poona: Deccan College Postgraduate and Research Institute, 1965.

———. *Horse Power: A History of the Horse and the Donkey in Human Societies*. London: Natural History Museum Publications, 1992.

———. "How Domestic Animals Have Shaped the Development of Human Societies." In *A Cultural History of Animals in Antiquity*, ed. by Linda Kalof, 71–96. New York: Berg, 2007.

———. "The Mammalian Remains from the Jericho Tell." *Proceedings of the Prehistoric Society* 45 (1979): 135–157.

———. *A Natural History of Domesticated Mammals*, 2nd ed. Cambridge: Cambridge University Press/ Natural History Museum, 1999.

———. "Origins of the Dog: Domestication and Early History." In *The Domestic Dog: Its Evolution, Behaviour, and Interactions with People*, ed. by James Serpell, 7–20. Cambridge: Cambridge University Press, 1995.

———. "Ritual Burial of a Dog and Six Domestic Donkeys." In *Excavations at Tell Brak: Nagar in the Third Millennium b.c.*, vol. 2, ed. by D. Oates, J. Oates, and H. McDonald, 327–338. Cambridge: British School of Archaeology in Iraq and the McDonald Institute for Archaeological Research, 2001.

———. "The Spread of Domestic Animals in Africa." In *The Archaeology of Africa: Food, Metals and Towns*, ed. by Thurstan Shaw, Paul Sinclair, Bassey Andah, and Alex Okpoko, 61–70. London: Routledge, 1993.

———. "The Unnatural World: Behavioural Aspects of Humans and Animals in the Process of Domestication." In *Animals and Human Society: Changing Perspectives*, ed. by Aubrey Manning and James Serpell, 23–35. New York: Routledge, 1994.

Clutton-Brock, J., K. Dennis-Bryan, P. L. Armitage, and P. A. Jewell. "Osteology of the Soay Sheep." *Bulletin of the British Museum (Natural History)(Zoology)* 55, no. 2 (1990): 1–56.

Clutton-Brock, Juliet, and Norman Hammond. "Hot Dogs: Comestible Canids in Preclassic Maya Culture at Cuello, Belize." *Journal of Archaeological Science* 21 (1994): 819–826.

Clutton-Brock, J., and N. Noe-Nygaard. "New Osteological and C-isotope Evidence on Mesolithic Dogs: Companions to Hunters and Fishers at Star Carr, Seamer Carr and Kongemose." *Journal of Archaeological Science* 17 (1990): 643–653.

Clutton-Brock, Tim, and Josephine Pemberton. *Soay Sheep Dynamics and Selection in an Island Population*. Cambridge: Cambridge University Press, 2004.

Cohen, Mark N. *The Food Crisis in Prehistory*. New Haven, Conn.: Yale University Press, 1977.

Columella, L. J. M. *Lucius Junius Moderatus Columella on Agriculture: De Re Rustica*, vol. 2, Loeb Classical Library. Trans. by E. S. Forster and E. H. Heffner. London: William Heinemann, and Cambridge, Mass.: Harvard University Press, 1968.

Corbett, L. K. "Morphological Comparisons of Australian and Thai Dingoes: A Reappraisal of Dingo Status, Distribution and Ancestry." *Proceedings of the Ecological Society of Australia* 13 (1985): 277–290.

Crane, Eva. "Honeybees." In *Evolution of Domesticated Animals*, ed. by I. L. Mason, 403–415. London: Longman, 1984.

Crawford, R. D. "Domestic Fowl." In *Evolution of Domesticated Animals*, ed. by I. L. Mason, 298–311. New York: Longman, 1984.

———. "Turkey." In *Evolution of Domesticated Animals*, ed. by I. L. Mason, 325–334. New York: Longman, 1984.

Crockford, Susan J. "A Commentary on Dog Evolution: Regional Variation, Breed Development and Hybridization with Wolves." In *Dogs through Time: An Archaeological Perspective*, ed. by Susan Crockford, 295–312. Oxford: BAR International Series 889, 2000.

Crosby, A. W. *The Columbian Exchange: Biological and Cultural Consequences of 1492*. Westport, Conn.: Greenwood Press, 1972.

———. *Ecological Imperialism: The Biological Expansion of Europe, 900–1900*. Cambridge: Cambridge University Press, 1986.

Cucchi, Thomas, and Jean-Denis Vigne. "Origin and Diffusion of the House Mouse in the Mediterranean." *Human Evolution* 21 (2006): 95–106.

Darwin, Charles. *On the Origin of Species by Means of Natural Selection or the Preservation of Favoured Races in the Struggle for Life*. London: John Murray, 1859.

———. *The Variation of Animals and Plants under Domestication*, 2nd ed. London: John Murray, 1890.

Davis, S. J. M. *The Archaeology of Animals*. London: Batsford, 1987.

Davis, S. J. M., and F. R. Valla. "Evidence for Domestication of the Dog 12,000 Years Ago in the Natufian of Israel." *Nature* 276, no. 5688 (1978): 608–610.

Dayan, Tamar. "Early Domesticated Dogs of the Near East." *Journal of Archaeological Science* 21 (1994): 633–640.

Deeben, Jos, and Nico Arts. "From Tundra Hunting to Forest Hunting: Later Upper Palaeolithic and Early Mesolithic." In *The Prehistory of the Netherlands*, vol. 1, ed. by L. P. Louwe Kooijmans et al., 151–153. Amsterdam: Amsterdam University Press, 2005.

de Waal, Frans. *The Ape and the Sushi Master: Cultural Reflections by a Primatologist*. New York: Penguin Books, 2001.

Dhavalikar, M. K. "From Farming to Pastoralism: Effects of Climatic Change in the Deccan." In *The Walking Larder: Patterns of Domestication, Pastoralism and Predation*, ed. by Juliet Clutton-Brock, 156–168. London: Unwin Hyman, 1989.

Diamond, Jared. *Guns, Germs and Steel: A Short History of Everybody for the Last 13,000 Years*. London: Vintage, 1998.

———. *The Third Chimpanzee: The Evolution and Future of the Human Animal*. New York: Harper Perennial, 2006.

Di Lernia, Savino, and Mauro Crernasc. "Taming Barbary Sheep: Wild Animal Management by Early Holocene Hunter-Gatherers at Uan Afuda (Libyan Sahara)." *Society of African Archaeologists: Nyame Akuma* 46 (December 1996): 43–54.

Dobney, K., and G. Larson, "Genetics and Animal Domestication: New Windows on an Elusive Process." *Journal of Zoology* 269 (2006): 261–271.

Driscoll, Carlos A., Juliet Clutton-Brock, Andrew C. Kitchener, and Stephen J. O'Brien. "The Taming of the Cat." *Scientific American* 300, no. 6 (June 2009): 56–63.

Driscoll, Carlos A., et al. "Near Eastern Origin of Cat Domestication." *Science* 317 (2007): 519–523.

Ellis, W. *Polynesian Researches*, 4 vols. London: Fisher, Jackson, 1839.

Epstein, H. *Domestic Animals of China*. Bucks, England: Commonwealth Agricultural Bureaux, Farnham Royal, 1969.

———. *The Origin of the Domestic Animals of Africa*, 2 vols. New York: Africana Publishing, 1971.

Epstein, H., and I. L. Mason. "Cattle." In *Evolution of Domesticated Animals*, ed. by I. L. Mason, 6–27. New York: Longman, 1984.

Eriksson, J., et al. "Identification of the *Yellow Skin* Gene Reveals a Hybrid Origin of the Domestic Chicken." *PLoS Genetics* 4, no. 2 (29 February 2008).

Evans-Pritchard, E. E. *The Nuer.* Oxford: Oxford University Press, 1940.

Evershed, Richard P., et al. "Earliest Date for Milk Use in the Near East and Southeastern Europe Linked to Cattle Herding." *Nature* 455 (25 September 2008): 528–531.

Fitts, R. L., C. Haselgrove, P. Lowther, and S. Willis. "Melsonby Revisited: Survey and Excavation 1992–95 at the Site of Discovery of the 'Stanwick' North Yorkshire Hoard of 1843." *Durham Archaeological Journal* 14–15 (1999): 1–52.

Foer, Jonathan Safran. *Eating Animals.* London: Hamish Hamilton, 2009.

Frazer, James George. *The Golden Bough: A Study in Magic and Religion*, 2 vols. London: Macmillan, 1911–1915.

Gallant, Johan. *The Story of the African Dog.* Pietermaritzburg: University of Natal Press, 2002.

Galton, Francis. "Domestication of Animals." In *Enquiries into Human Faculty and Its Development*, 2nd ed., 173–193. London: Dent and Dutton, Everyman, 1907.

Garnery, L., J.-M. Cornuet, and M. Solignac. "Evolutionary History of the Honey Bee *Apis mellifera* Inferred from Mitochondrial DNA Analysis." *Molecular Ecology* 1, no. 3 (28 June 2008): 145–154.

Garrard, Andrew, Susan Colledge, and Louise Martin. "The Emergence of Crop Cultivation and Caprine Herding in the 'Marginal Zone' of the Southern Levant." In *The Origins and Spread of Agriculture and Pastoralism in Eurasia*, ed. by David R. Harris, 208–221. London: UCL Press, 1996.

Gauthier-Pilter, Hilde, and Anne Innis Dagg. *The Camel: Its Evolution, Ecology, Behavior and Relationship to Man.* Chicago: University of Chicago Press, 1981.

Gautier, Achilles. "What's in a Name." In *Skeletons in Her Cupboard: Festschrift for Juliet Clutton-Brock*, ed. by Anneke Clason, Sebastian Payne, and Hans-Peter Uerpmann, 91–98. Oxford: Oxbow Monograph 34, 1993.

Gentry, Anthea, Juliet Clutton-Brock, and Colin Groves. "The Naming of Wild Animal Species and Their Domestic Derivatives." *Journal of Archaeological Science* 31 (2004): 645–651.

Germonpréa, Mietje, et al. "Fossil Dogs and Wolves from Palaeolithic Sites in Belgium, the Ukraine and Russia: Osteometry, Ancient DNA and Stable Isotopes." *Journal of Archaeological Science* 36, no. 2 (2009): 473–490.

Goring-Morris, Nigel, and Liora Kolska Horwitz. "Funerals and Feasts during the Pre-Pottery Neolithic B of the Near East." *Antiquity* 81 (2007): 902–919.

Grahame, Iain. "Peafowl." In *Evolution of Domesticated Animals*, ed. by I. L. Mason, 315–321. New York: Longman, 1984.

Grigson, Caroline. "The Craniology and Relationships of Four Species of *Bos* 5. *Bos indicus* L." *Journal of Archaeological Science* 7 (1980): 3–32.

———. "The Earliest Domestic Horses in the Levant? New Finds from the Fourth Millennium of the Negev." *Journal of Archaeological Science* 20 (1993): 645–655.

Groves, Colin P. *Ancestors for the Pigs.* Technical Bulletin No 3, Department of Prehistory, Research School of Pacific Studies, Australian National University, 1981.

———. "Current Views on Taxonomy and Zoogeography of the Genus *Sus*." In *Pigs and Humans: 10,000 Years of Interaction*, ed. by Umberto Albarella, Keith Dobney, Anton Ervynck, and Peter Rowley-Conwy, 18–29. Oxford: Oxford University Press, 2007.

———. "The Taxonomy, Distribution, and Adaptation of Recent Equids." In *Equids in the Ancient World*, ed. by Richard H. Meadow and Hans-Peter Uerpmann, 1: 27–32. Wiesbaden: Dr Ludwig Reichart, 1986.

Guppy, Nicholas. *Wai-Wai through the Forests North of the Amazon.* London: Penguin Books, 1961.

Halstead, Paul. "Sheep in the Garden: The Integration of Crop and Livestock Husbandry in Early

Farming Regimes of Greece and Southern Europe." In *Animals in the Neolithic of Britain and Europe*, ed. by Dale Serjeantson and David Field, 42–55. Oxford: Oxbow Books, 2006.

Harris, David R., ed. *The Origins and Spread of Agriculture and Pastoralism in Eurasia*. London: UCL Press, 1996.

Harting, James E. *The Birds of Shakespeare*. 1871. Repr., Chicago: Argonaut, 1965.

Haynes, Gary, ed. *American Megafaunal Extinctions at the End of the Pleistocene*. New York: Springer, 2009.

Hedges, Simon, and Donny Gunaryadi. "Reducing Human-Elephant Conflict: Do Chillies Help Deter Elephants from Entering Crop Fields?" *Oryx* 44, no. 1 (2009): 133–138.

Hemmer, Helmut. *Domestication: The Decline of Environmental Appreciation*. Cambridge: Cambridge University Press, 1990.

Herodotus. *The Histories of Herodotus*, vol. 1. Trans. by George Rawlinson. London: Dent, Everyman's Library, 1970.

Hodder, Ian. *Çatalhöyük: The Leopard's Tale, Revealing the Mysteries of Turkey's Ancient "Town."* London: Thames and Hudson, 2006.

Homer. *Iliad*. Trans. by Martin Hammond. London: Penguin Books, 1987.

———. *Odyssey*. Trans. by E. V. Rieu. London: Penguin Books, 1946.

Horwitz, Liora Kolska, and Gila Kahila Bar-Gal. "The Origin and Genetic Status of Insular Caprines in the Eastern Mediterranean: A Case Study of Free-Ranging Goats (*Capra aegagrus cretica*) on Crete." *Human Evolution* 21 (2006): 123–138.

Houlihan, Patrick F. *The Animal World of the Pharaohs*. New York: Thames and Hudson, 1996.

———. *The Birds of Ancient Egypt*. Warminster, England: Aris and Phillips, 1986.

Hrdy, Sarah Blaffer. *Mothers and Others: The Evolutionary Origins of Mutual Understanding*. Cambridge, Mass.: Harvard University Press, 2009.

Huang, Yong-fu, Xian-wei Shi, and Ya-ping Zhang. "Mitochondrial Genetic Variation in Chinese Pigs and Wild Boars." *Biochemical Genetics* 37, nos. 11/12 (1999): 335–343.

Huxley, Thomas Henry. *Man's Place in Nature*. London: Dent and Dutton, Everyman, 1906.

Ingold, T. "From Trust to Domination: An Alternative History of Human-Animal Relations." In *Animals and Human Society: Changing Perspectives*, ed. by A. Manning and J. Serpell. New York: Routledge, 1994.

———. *Hunters, Pastoralists and Ranchers: Reindeer Economies and Their Transformations*. Cambridge: Cambridge University Press, 1980.

International Commission on Zoological Nomenclature. "Usage of 17 Specific Names Based on Wild Species Which Are Predated by or Contemporary with Those Based on Domestic Animals (Lepidoptera, Osteichthyes, Mammalia): Conserved." *Bulletin of Zoological Nomenclature* 60, no. 1 (2003): 81–84.

Ishiguro, Naotaka, Naohiko Okumura, Akira Matsui, and Nobuo Shigehara. "Molecular Genetic Analysis of Ancient Japanese Dogs." In *Dogs through Time: An Archaeological Perspective*, ed. by Susan Crockford, 287–292. Oxford: BAR International Series 889, 2000.

Itan, Yuval, Adam Powell, Mark A. Beaumont, Joachim Burger, and Mark G. Thomas. "The Origins of Lactase Persistence in Europe." *PLoS Computational Biology* 5, no. 8 (2009): 1–13.

Jansen, Thomas, et al. "Mitochondrial DNA and the Origins of the Domestic Horse." *Proceedings of the National Academy of Sciences* 99, no. 16 (August 2002): 10905–10910.

Jewell, Rebecca Tamsin. "Understanding Pacific Feather Artefacts through Drawing." Ph.D. diss., Royal College of Art, 2004.

Jin, X-B., and A. L. Yen. "Conservation and the Cricket Culture in China." *Journal of Insect Conservation* 2 (1998): 211–216.

Kadwell, Miranda, et al. "Genetic Analysis Reveals the Wild Ancestors of the Llama and the Alpaca." *Proceedings of the Royal Society* B 268 (2001): 2575–2584.

Kayser, Manfred. "The Human Genetic History of Oceania: Near and Remote Views of Dispersal." *Current Biology* 20, no. 4 (February 2010): 202–207.

Kenyon, K. M. *Digging Up Jericho.* London: Ernest Benn, 1957.

Khazanov, A. M. *Nomads and the Outside World.* Trans. by Julia Crookenden. Cambridge: Cambridge University Press, 1984.

King, J. E. "Mammal Bones from Khirokitia and Erimi." In *Khirokitia: Final Report on the Excavation of a Neolithic Settlement in Cyprus on Behalf of the Department of Antiquities, 1936–1946,* by P. Dikaios, 431–437. Oxford: Oxford University Press, 1953.

Kingsley, Charles. *The Water Babies.* London: Macmillan, 1910.

Klein, Richard. "The Prehistory of Stone Age Herders in the Cape Province of South Africa." In *Prehistoric Pastoralism in Southern Africa*, ed. by Martin Hall and Andrew B. Smith. South African Archaeological Society, Goodwin Series, vol. 5, June 1986, 5–12.

Koler-Matznick, Janice, I. Lehr Brisbin Jr., Mark Feinstein, and Susan Bulmer. "An Updated Description of the New Guinea Singing Dog (*Canis hallstromi,* Troughton 1957)." *Journal of Zoology* 261, no. 2 (2003): 109–118.

Komiyama, Tomoyoshi, Hiroyuki Kobayashia, Yoshio Tatenob, Hidetoshi Inokoa, Takashi Gojoborib, and Kazuho Ikeob. "An Evolutionary Origin and Selection Process of Goldfish." *Gene* 430, nos. 1–2 (1 February 2009): 5–11.

Koop, Ben F., and Susan J. Crockford. "Ancient DNA Evidence of a Separate Origin for North American Indigenous Dogs." In *Dogs through Time: An Archaeological Perspective*, ed. by Susan Jane Crockford, 271–284. Oxford: BAR International Series 889, 2000.

Kopper, J. S., and W. H. Waldren. "Balearic Prehistory: A New Perspective." *Archaeology* 20 (1967): 108–115.

Lane Fox, Robin. *Travelling Heroes: Greeks and Their Myths in the Epic Age of Homer.* London: Allen Lane, 2008.

Larson, Greger, Umberto Albarella, Keith Dobney, and Peter Rowley-Conwy. "Current Views on *Sus* Phylogeography and Pig Domestication as Seen through Modern mtDNA Studies." In *Pigs and Humans: 10,000 Years of Interaction*, ed. by Umberto Albarella, Keith Dobney, Anton Ervynck, and Peter Rowley-Conwy, 30–41. Oxford: Oxford University Press, 2007.

Larson, Greger, et al. "Patterns of East Asian Pig Domestication, Migration, and Turnover Revealed by Modern and Ancient DNA." *Proceedings of the National Academy of Sciences* 107, no. 17 (27 April 2010): 7686–7691.

Larson, Greger, et al. "Phylogeny and Ancient DNA of *Sus* Provides Insights into Neolithic Expansion in Southeast Asia and Oceania." *Proceedings of the National Academy of Sciences* 104, no. 12 (20 March 2007): 4834–4839.

Larson, Greger, et al. "Worldwide Phylogeography of Wild Boar Reveals Multiple Centres of Pig Domestication." *Science* 307 (2005): 1618–1621.

Leonard, J. A., Robert K. Wayne, Jane Wheeler, Raúl Valadez, Sonia Guillén, and Carles Vilà. "Ancient DNA Evidence for Old World Origin of New World Dogs." *Science* 298 (2002): 1613–1616.

Lernia, Savino Di, and Mauro Crernasch. "Taming Barbary Sheep: Wild Animal Management by Early Holocene Hunter-Gatherers at Uan Afuda (Libyan Sahara)." *Society of African Archaeologists: Nyame Akuma* 46 (1996): 43–54.

Lever, Christopher. *Naturalized Mammals of the World.* London: Longman, 1985.

Levine, Marsha, Colin Renfrew, and Katie Boyle. *Prehistoric Steppe Adaptation and the Horse.* Cambridge: McDonald Institute for Archaeological Research, 2003.

Lhote, Henri. *The Search for the Tassili Frescoes.* Trans. by Alan Houghton Brodrick. London: Hutchinson, 1959.

Linnaei, Caroli. *Systema Naturae, 1758.* Facsimile ed. London: Trustees of the British Museum (Natural History), 1958.

Lipinski, Monika J., et al. "The Ascent of Cat Breeds: Genetic Evaluations of Breeds and Worldwide Random Bred Populations." *Genomics* 91, no. 1 (2008): 12–21.

Littauer, M. A., and J. H. Crouwel. *Wheeled Vehicles and Ridden Animals in the Ancient Near East.* Leiden/Köln: E. J. Brill, 1979.

Liu, Y. P., et al. "Multiple Maternal Origins of Chickens: Out of the Asian Jungles." *Molecular Phylogenetics and Evolution* 38, no. 1 (2006): 12–19.

Livy. *The War with Hannibal.* Trans. by Aubrey de Sélincourt. London: Penguin Books, 1965.

Loftus, Ronan, and Patrick Cunningham. "Molecular Genetic Analysis of African Zeboid Populations." In *The Origins and Development of African Livestock: Genetics, Linguistics and Ethnography*, ed. by Roger M. Blench and Kevin C MacDonald, 251–258. London: UCL Press, 2000.

Loftus, R. T., et al. "Evidence for Two Independent Domestications of Cattle." *Proceedings of the National Academy of Sciences,* 91, no 7 (29 March 1994): 2757–2761.

Ludwig, Arne, et al. "Coat Color Variation at the Beginning of Horse Domestication." *Science* 324 (24 April 2009): 485.

Luikart, G., H. Fernandez, M. Mashkour, P. R. England, and P. Taberlet. "Origins and Diffusion of Domestic Goats Inferred from DNA Markers: Example Analyses of mtDNA, Y Chromosome, and Microsatellites." In *Documenting Domestication: New Genetic and Archaeological Paradigms*, ed. by Melinda A. Zeder et al., 294–305. Berkeley: University of California Press, 2006.

Lumholz, C. *Among Cannibals.* London: J. Murray, 1899.

Mack, Arien, ed. *Humans and Other Animals.* Columbus: Ohio State University Press, 1995.

Marra, Antonella Cinzia. "Pleistocene Mammals of Mediterranean Islands." *Quaternary International* 129 (2005): 5–14.

Mayer, John J., and I. Lehr Brisbin Jr. *Wild Pigs of the United States: Their History, Morphology, and Current Status.* Athens: University of Georgia Press, 1991.

Mayor, Adrienne. *The First Fossil Hunters: Paleontology in Greek and Roman Times.* Princeton, N.J.: Princeton University Press, 2000.

Mayr, Ernst. *Animal Species and Evolution.* Cambridge, Mass.: Belknap, 1966.

Meadow, Richard H. "The Origins and Spread of Agriculture and Pastoralism in Northwestern South Asia." In *The Origins and Spread of Agriculture and Pastoralism in Eurasia*, ed. by David R. Harris, 402–406. London: UCL Press, 1996.

Meggitt, M. "Australian Aborigines and Dingoes." In *Man, Culture and Animals*, ed. by A. Leeds and P. Vayda, 7–26. American Association for the Advancement of Science Symposium, 1965.

Meghen, Ciaran, David MacHugh, B. Sauveroche, G. Kana, and Dan Bradley. "Characterization of the Kuri Cattle of Lake Chad Using Molecular Genetic Techniques." In *The Origins and Development of African Livestock: Genetics, Linguistics and Ethnography*, ed. by Roger M. Blench and Kevin C. MacDonald, 259–268. London: UCL Press, 2000.

Mellaart, James. *Çatal Hüyük: A Neolithic Town in Anatolia.* London: Thames and Hudson, 1967.

———. *The Neolithic of the Near East.* London: Thames and Hudson, 1975.

Mellars, Paul A. "The Palaeolithic and Mesolithic." In *British Prehistory: A New Outline*, ed. by Colin Renfrew, 77–81. London: Duckworth, 1974.

Melville, Elinor. *A Plague of Sheep: Environmental Consequences of the Conquest of Mexico.* Cambridge: Cambridge University Press, 1997.

Mengoni Goñalons, G. L., and H. D. Yacobaccio. "The Domestication of South American Camelids a View from the South-Central Andes." In *Documenting Domestication: New Genetic and Archaeological Paradigms*, ed. by M. A. Zeder et al., 279–293. Berkeley: University of California Press, 2006.

Mithen, Steven. *After the Ice: A Global Human History, 20,000–5000 b.c.* London: Weidenfeld and Nicolsen, 2003.

Morey, D., and M. D. Wiant. "Early Holocene Domestic Dog Burials from the North American Midwest." *Current Anthropology* 33 (1992): 224–229.

Morwood, M. J., et al. "Archaeology and Age of a New Hominin from Flores in Eastern Indonesia." *Nature* 431 (28 October 2004): 1087–1091.

Müller-Haye, B. "Guinea-pig or Cuy." In *Evolution of Domesticated Animals*, ed. by I. L. Mason, 252–257. New York: Longman, 1984.

Musil, Rudolf. "Domestication of Wolves in Central European Magdalenian Sites." In *Dogs through Time: An Archaeological Perspective*, ed. by Susan Janet Crockford, 21–28. Oxford: BAR International Series 889, 2000.

Muzzolini, Alfred. *L'Art Rupestre Préhistorique des Massifs Centraux Sahariens*. Oxford: Cambridge Monographs in African Archaeology 16, BAR International Series 318, 1986.

———. "The Emergence of a Food-Producing Economy in the Sahara." In *The Archaeology of Africa: Food, Metals and Towns*, ed. by Thurstan Shaw, Paul Sinclair, Bassey Andah, and Alex Okpoko, 227–239. New York: Routledge, 1993.

Nakajima, Sadahiko, Mariko Yamamoto, and Natsumi Yoshimoto. 2009. "Dogs Look Like Their Owners: Replications with Racially Homogenous Owner Portraits." *Anthrozoös* 22, no. 2 (2009): 173–181.

Oakley, Kenneth Page. *Man the Toolmaker*. Chicago: University of Chicago Press, 1968.

Oates, Joan. "A Note on the Early Evidence for Horse and the Riding of Equids in Western Asia." In *Prehistoric Steppe Adaptation and the Horse*, ed. by Marsha Levine, Colin Renfrew, and Katie Boyle, 115–125. Cambridge: McDonald Institute for Archaeological Research, 2003.

Olsen, Sandra L. "Early Horse Domestication on the Eurasian Steppe." In *Documenting Domestication: New Genetic and Archaeological Paradigms*, ed. by Melinda A. Zeder et al., 251–253. Berkeley: University of California Press, 2006.

O'Rourke, Dennis H., and Jennifer A. Raff. "The Human Genetic History of the Americas: The Final Frontier." *Current Biology* 20 (23 February 2010): R202–R207.

Outram, Alan K., et al. "The Earliest Horse Harnessing and Milking." *Science* 323, no. 5919 (6 March 2009): 1332–1335.

Pang, Jun-Feng, et al. "mtDNA Data Indicate a Single Origin for Dogs South of Yangtze River, Less Than 16,300 Years Ago, from Numerous Wolves." *Molecular Biology and Evolution* 26, no. 12 (2009): 2849–2864.

Parr, Peter. "The Levant in the Early First Millennium b.c." In *The Cambridge Encyclopedia of Archaeology*, ed. by Andrew Sherratt and Grahame Clark, 196–199. Cambridge: Cambridge University Press, 1980.

Parwanker, Seema J. *Man and Animal Relationships in Early Farming Communities of Western India with Special Reference to Inamgaon*. Oxford: BAR, S, 1639, 2007.

Pharr, Clyde, trans. *The Theodosian Code and Novels and the Sirmondian Constitutions*. Princeton, N.J.: Princeton University Press, 1952.

Phillipson, David. *African Archaeology*, 2nd ed. Cambridge: Cambridge University Press, 1993.

Piggott, Stuart. *The Earliest Wheeled Transport from the Atlantic Coast to the Caspian Sea*. London: Thames and Hudson, 1983.

Pliny. *The Natural History of Pliny*, 6 vols. Trans. by John Bostock. London: Henry G. Bohn, 1855.

Podberscek, Anthony L. "Good to Pet and Eat: The Keeping and Consuming of Dogs and Cats in South Korea." *Journal of Social Issues* 65, no. 3 (2009): 617.

Polo, Marco. *The Travels of Marco Polo*. Trans. by Ronald Latham. London: Penguin Books, 1988.

Polybius. *The Rise of the Roman Empire*. Trans. by Ian Scott-Kilvert. London: Penguin Books, 1979.

Poplin, F. "Origine du mouflon de Corse dans une nouvelle perspective paléontologique: par marronage." *Annales Génétique et de Sélection Animale* 11, no. 2 (1979): 133–143.

Porter, Valerie. *Cattle: A Handbook to the Breeds of the World.* London: Christopher Helm, 1991.

Postgate, Nicholas. "The Assyrian Empire." In *The Cambridge Encyclopedia of Archaeology*, ed. by Andrew Sherratt and Grahame Clark, 186–192. Cambridge: Cambridge University Press, 1980.

———. "The Equids of Sumer Again." In *Equids in the Ancient World*, vol. 1, ed. by Richard H. Meadow and Hans-Peter Uerpmann, 194–206. Wiesbaden: Dr Ludwig Reichert, 1986.

———. *The First Empires*. Oxford: Elsevier-Phaidon, 1977.

Raulwing, Peter, and Juliet Clutton-Brock. "The Buhen Horse: Fifty Years after Its Discovery (1958–2008)." *Journal of Egyptian History* 2, nos. 1–2 (2009): 1–106.

Renfrew, Colin. "Archaeogenetics: Towards a 'New Synthesis.'" *Current Biology* 20 (23 February 2010): R162–R165.

Rice, Michael. *The Power of the Bull.* New York: Routledge, 1998.

———. *Swifter Than the Arrow: The Golden Hunting Hounds of Ancient Egypt.* London: I. B. Taurus, 2006.

Robertshaw, Peter. "The Beginnings of Food Production in Southwestern Kenya." In *The Archaeology of Africa: Food, Metals and Towns*, ed. by Thurstan Shaw, Paul Sinclair, Bassey Andah, and Alex Okpoko, 359–362. London: Routledge, 1993.

Rudenko, S. I. *Frozen Tombs of Siberia: The Pazyryk Burials of Iron-Age Horsemen.* London: Dent, 1970.

———. *Sibirskaya kollektsiya Petra I* [The Siberian Collection of Peter the Great]. SAI, vyp. D319. Moscow, Leningrad, 1962.

Ryan, Kathleen, Karega Mumene, Samuel M. Kahinju, and Paul N. Kunoni. "Ethnographic Perspectives on Cattle Management in Semi-arid Environments: A Case Study from Maasailand." In *The Origins and Development of African Livestock: Genetics, Linguistics and Ethnography*, ed. by Roger M. Blench and Kevin C. MacDonald, 462–477. London: UCL Press, 2000.

Sablin, M. V., and G. A. Khlopachev. "The Earliest Ice Age Dogs: Evidence from Eliseevichi." *Current Anthropology* 43 (2002): 795–799.

Sankalia, Hasmukh Dhirajlal. *Excavations at Langhnaj: 1944–63 Part I: Archaeology*. Poona: Deccan College Postgraduate and Research Institute, 1965.

Savolainen, Peter, Thomas Leitner, Alan N. Wilton, Elizabeth Matisoo-Smith, and Joakim Lundeberg. "A Detailed Picture of the Origin of the Australian Dingo, Obtained from the Study of Mitochondrial DNA." *Proceedings of the National Academy of Sciences* 101, no. 33 (17 August 2004): 12387–12390.

Savolainen, Peter, Ya-ping Zhang, Jing Luo, Joakim Lundeberg, and Thomas Leitner. "Genetic Evidence for an East Asian Origin of Domestic Dogs." *Science* 298 (22 November 2002): 1610–1613.

Schulting, R., A. Tresset, and C. Dupont. "From Harvesting the Sea to Stock Rearing Along the Atlantic Facade of North-West Europe." *Environmental Archaeology* 9 (2004): 131–142.

Schulting, Rick J., and Michael P. Richards. "Dogs, Divers, Deer and Diet: Stable Isotope Results from Star Carr and a Response to Dark." *Journal of Archaeological Science* 36 (2009): 498–503.

Schuster, Stephen C., et al. "Complete Khoisan and Bantu Genomes from Southern Africa." *Nature* 463 (18 February 2010): 943–947.

Schwartz, Marion. *A History of Dogs in the Early Americas.* New Haven, Conn.: Yale University Press, 1997.

Sherratt, Andrew. "Plough and Pastoralism: Aspects of the Secondary Products Revolution." In *Pattern of the Past: Studies in Honour of David Clarke*, ed. by I. Hodder, G. Isaac, and N. Hammond, 261–305. Cambridge: Cambridge University Press, 1981.

Siculus, Diodorus. *Diodorus Siculus*, Loeb Classical Library, vols. 1–6. Trans. by C. H. Oldfather. Cambridge, Mass.: Harvard University Press, 1989.

Simmons, Alan H. *Faunal Extinction in an Island Society: Pygmy Hippopotamus Hunters of Cyprus.* New York: Springer, 1999.

Simoons, F. J. "Gayal or Mithan." In *Evolution of Domesticated Animals*, ed. by I. L. Mason, 34–39. New York: Longman, 1984.

Sittert, Lance van, and Sandra Swart, eds. *Canis Africanis: A Dog History of Southern Africa.* Boston: Brill, 2008.

Smith, Andrew B. "The Origins of the Domesticated Animals of Southern Africa." In *The Origins and Development of African Livestock: Genetics, Linguistics and Ethnography*, ed. by Roger M. Blench and Kevin C. MacDonald, 222–238. London: UCL Press, 2000.

———. *Pastoralism in Africa: Origins and Development Ecology.* London: Hurst, 1992.

Smith, Bradley P., and Carla A. Litchfield. "A Review of the Relationship between Indigenous Australians, Dingoes (*Canis dingo*) and Domestic Dogs (*Canis familiaris*)." *Anthrozoös* 22, no. 2 (2009): 111–128.

Speller, Camilla F., et al. "Ancient Mitochondrial DNA Analysis Reveals Complexity of Indigenous North American Turkey Domestication." *Proceedings of the National Academy of Sciences* 107, no. 7 (16 February 2010): 2807–2812.

Spotorno, Angel E., J. C. Marin, G. Manriquez, J. P. Valladares, E. Rico, and C. Rivas. "Ancient and Modern Steps during the Domestication of Guinea-Pigs." *Journal of Zoology* 270, no. 1 (2006): 57–62.

Street, Martin. "Ein frühmesolithischer Hund und Hundeverbiß an Knochen vom Fundplatz Bedburg-Königshoven, Niederrhein." *Archäologische Informationen* 12 (1989): 203–215.

Sutcliffe, Antony J. *On the Track of Ice Age Mammals.* London: British Museum (Natural History), 1985.

Tani, Yutaka. "The Geographical Distribution and Function of Sheep Flock Leaders: A Cultural Aspect of the Man-Domesticated Animal Relationship in Southwestern Eurasia." In *The Walking Larder: Patterns of Domestication, Pastoralism and Predation*, ed. by Juliet Clutton-Brock, 185–199. London: Unwin Hyman: 1990.

Tchernov, Eitan, and François F. Valla. "Two New Dogs, and Other Natufian Dogs, from the Southern Levant." *Journal of Archaeological Science* 24, no. 1 (1997): 65–95.

Thomas, Elizabeth Marshall. *The Old Way: A Story of the First People.* New York: Picador, 2007.

Thomas, Julian. "The Cultural Context of the First Use of Domesticates in Continental Central and Northwest Europe." In *The Origins and Spread of Agriculture and Pastoralism in Eurasia*, ed. by David R. Harris. London: UCL Press, 1996.

Thwaite, A. *Edmund Gosse: A Literary Landscape.* Stroud, Gloucestershire: Tempus, 2007.

Tisdell, C. A. *Wild Pigs: Environmental Pest or Economic Resource?* Oxford: Pergamon Press, 1982.

Tishkoff, Sarah A., et al. "Convergent Adaptation of Human Lactase Persistence in Africa and Europe." *Nature Genetics* 39 (2007): 31–40.

Titcomb, Margaret. *Dog and Man in the Ancient Pacific with Special Attention to Hawaii.* Bernice P. Bishop Museum Special Publication 59, Honolulu, Hawaii, 1969.

Tristram, H. B. *The Natural History of the Bible*. London: Society for Promoting Christian Knowledge, 1889.

Troughton, Ellis. "A New Native Dog from the Papuan Highlands." *Proceedings of the Royal Zoological Society of New South Wales, 1955–56* (8 May 1957): 93–94.

Turner, Elaine. "Results of a Recent Analysis of Horse Remains Dating to the Magdalenian Period at Solutré, France." In *Equids in Time and Space: Papers in Honour of Vera Eisenmann*, ed. by Marjan Mashkour, 70–89. Oxford: Oxbow Books, 2006.

Uerpmann, Hans-Peter. *The Ancient Distribution of Ungulate Mammals in the Middle East*. Wiesbaden: Dr. Ludwig Reichert, 1987.

———. "Animal Domestication: Accident or Intention?" In *The Origins and Spread of Agriculture and Pastoralism in Eurasia*, ed. by David R. Harris, 227–237. London: UCL Press, 1996.

Vainshtein, Sevyan. *Nomads of South Siberia: The Pastoral Economies of Tuva*. Trans. by Michael Colenso. Cambridge: Cambridge University Press, 1980.

Valla, E. F. F. "Two New Dogs, and Other Natufian Dogs, from the Southern Levant." *Journal of Archaeological Science* 24, no, 1 (1997): 65–95.

Vigne, Jean-Denis. "Zooarchaeological Aspects of the Neolithic Diet Transition in the Near East and Europe and Their Putative Relationships with the Neolithic Demographic Transition." In *The Neolithic Demographic Transition and Its Consequences*, ed. by J.-P. Bocquet Appel and O. Bar-Yosef, 179–205. New York: Springer Verlag, 2008.

Vigne, J.-D., J. Guilaine, K. Debue, L. Haye, and P. Gérard. "Early Taming of the Cat in Cyprus." *Science* (Brevia) 304, no. 5668 (2004): 259.

Vilà, Carles, Jennifer A. Leonard, and Albano Beja-Pereira. "Genetic Documentation of Horse and Donkey Domestication." In *Documenting Domestication: New Genetic and Archaeological Paradigms*, ed. by Melinda A. Zeder et al., 342–354. Berkeley: University of California Press, 2006.

Vilà, C., et al. "Multiple and Ancient Origins of the Dog." *Science* 276 (13 June 1997): 1687–1689.

Vinnicombe, Patricia. *People of the Eland: Rock Paintings of the Drakensberg Bushmen as a Reflection of Their Life and Thought*. Pietermaritzburg: University of Natal Press, 1976.

Vitebsky, Piers. *Reindeer People*. London: Harper Perennial, 2005.

von Holdt, Bridgett M., et al. "Genome-Wide SNP and Haplotype Analyses Reveal a Rich History Underlying Dog Domestication." *Nature* 464 (8 April 2010): 898–902.

Washburn, Sherwood L., and C. S. Lancaster. "The Evolution of Hunting." In *Man the Hunter*, ed. by Richard B. Lee and Irven Devore, 293–303. Chicago: Aldine, 1968.

Wayne, Robert K., Jennifer A. Leonard, and Carles Vilà. "Genetic Analysis of Dog Domestication." In *Documenting Domestication: New Genetic and Archaeological Paradigms*, ed. by Melinda A. Zeder et al., 279–293. Berkeley: University of California Press, 2006.

Weir, Barbara. "Notes on the Origin of the Domestic Guinea-pig." In *The Biology of Hystricomorph Rodents*, ed. by I. W. Rowlands and Barbara J. Weir, 437–446. Symposia of the Zoological Society of London 34 (1974).

Wendorf, Fred, and Romuald Schild. "Nabta Playa and Its Role in Northeast African Prehistory." *Journal of Anthropological Archaeology* 17 (1998): 97–123.

West, Barbara, and Ben-Xiong Zhou. "Did Chickens Go North? New Evidence for Domestication." *World's Poultry Science Journal* 45 (1989): 205–218.

Wheeler, J. C. "Evolution and Present Situation of the South American Camelidae." *Biological Journal of the Linnean Society* 54 (1995): 271–295.

———. "On the Origin and Early Development of Camelid Pastoralism in the Andes." In *Animals and Archaeology 3: Early Hunters and Their Flocks*, ed. by J. Clutton-Brock and C. Grigson, 395–410. Oxford: BAR International Series, 202, 1984.

Wheeler, J. C., A. J. F. Russel, and H. Redden. "Llamas and Alpacas: Pre-conquest Breeds and Post-conquest Hybrids." *Journal of Archaeological Science* 22, no. 6 (1995): 833–840.

Woodman, P. C. *The Mesolithic in Ireland*. Oxford: BAR British Series 58, 1978.

Xenophon. *Xenophon's Minor Works*. Trans. by J. S. Watson. London: George Bell, 1884.

———. *The Persian Expedition*. Trans. by Rex Warner. London: Penguin Books, 1975.

Zeder, Melinda A., Daniel G. Bradley, Eve Emshwiller, and Bruce D. Smith, eds. *Documenting Domestication: New Genetic and Archaeological Paradigms*. Berkeley: University of California Press, 2006.

Zeuner, F. E. *A History of Domesticated Animals*. London: Hutchinson, 1963.

Zhigunov, P. S., ed. *Reindeer Husbandry*. Jerusalem: Israel Program for Scientific Translations, 1968.

Zhon-ge, Zhang. "Goldfish." In *Evolution of Domesticated Animals*, ed. by I. L. Mason, 381–385. New York: Longman, 1984.

Index